Radical and Ion-pairing Strategies in
Asymmetric Organocatalysis

Green Chemistry and Organocatalysts Set

coordinated by
Max Malacria and Géraldine Masson

Radical and Ion-pairing Strategies in Asymmetric Organocatalysis

Maxime R. Vitale
Sylvain Oudeyer
Vincent Levacher
Jean-François Brière

ELSEVIER

First published 2017 in Great Britain and the United States by ISTE Press Ltd and Elsevier Ltd

ISTE Press Ltd
27-37 St George's Road
London SW19 4EU
UK

Elsevier Ltd
The Boulevard, Langford Lane
Kidlington, Oxford, OX5 1GB
UK

www.iste.co.uk

www.elsevier.com

Notices

Knowledge and best practice in this field are constantly changing. As new research and experience broaden our understanding, changes in research methods, professional practices, or medical treatment may become necessary.

Practitioners and researchers must always rely on their own experience and knowledge in evaluating and using any information, methods, compounds, or experiments described herein. In using such information or methods they should be mindful of their own safety and the safety of others, including parties for whom they have a professional responsibility.

To the fullest extent of the law, neither the Publisher nor the authors, contributors, or editors, assume any liability for any injury and/or damage to persons or property as a matter of products liability, negligence or otherwise, or from any use or operation of any methods, products, instructions, or ideas contained in the material herein.

For information on all our publications visit our website at http://store.elsevier.com/

British Library Cataloguing-in-Publication Data
A CIP record for this book is available from the British Library
Library of Congress Cataloging in Publication Data
A catalog record for this book is available from the Library of Congress
ISBN 978-1-78548-127-7

Printed and bound in the UK and US

Contents

Foreword vii
Luc NEUVILLE and Géraldine MASSON

**Chapter 1. SOMO and Photoredox Asymmetric
Organocatalysis** 1
Maxime R. VITALE

    1.1. Introduction. 1
    1.2. Asymmetric SOMO organocatalysis 2
        1.2.1. The concept of SOMO organocatalysis 2
        1.2.2. Enantioselective α-functionalization of
    carbonyl compounds 6
        1.2.3. Enantioselective cascade and pseudo-cycloaddition
    processes 18
        1.2.4. Conclusion. 23
    1.3. Asymmetric photoredox organocatalysis 23
        1.3.1. Photoredox catalysis 24
        1.3.2. Synthetic applications of photoredox organocatalysis 27
    1.4. Conclusions. 72
    1.5. Bibliography 72

**Chapter 2. Chiral Quaternary Ammonium Salts in
Organocatalysis** 87
Sylvain OUDEYER, Vincent LEVACHER and Jean-François BRIÈRE

    2.1. Introduction. 87
    2.2. Phase transfer catalysis 91
        2.2.1. Phase transfer catalyst: properties and mechanism 91
        2.2.2. Chiral catalysts: an overview 96
        2.2.3. C–C bond formation 102

2.2.4. C–O bond formation   112
2.2.5. C–N bond formation   120
2.2.6. C–Y bond formation (Y = S, X...).   125
2.2.7. New developments in PTC   130
2.3. Cooperative ion pairs organocatalysts   132
2.3.1. Mechanisms of action   134
2.3.2. Reactions based upon type 1 mechanism   136
2.3.3. Reactions based upon type 2 mechanism   149
2.4. Conclusion   152
2.5. Bibliography   153

Index   175

Foreword

Organocatalysis, as opposed to [metalo]-catalysis, is a process promoted purely by organic molecules, and has a long history, for example occurring extensively in biological systems. Although long known, the term "Organocatalysis", was only introduced in 2000, and has since then stimulated an explosion of new research in organic chemistry resulting in major advances, especially in the field of asymmetric synthesis. The types of catalyst used in this area are extremely diverse, including for example amines, urea, acids, alcohols, halogenated species and carbenes, among others, and offer a large panel of bond formation providing powerful tools for the construction of complex chiral units.

With this series of books, our intention is to bring together all important aspects of the field of asymmetric organocatalysis and to present them in a format that is most useful to a wide range of scientists, including students of chemistry, expert practitioners, and chemists contemplating the possibility of using an asymmetric organocatalytic reaction in their own research.

Chapter 1 is centered on new or recent (re)-emerging specific activation mode of particular functional groups through SOMO (Singly Occupied Molecular Orbital) and photoredox asymmetric organocatalysis. Both processes, relying on discreet radical pathways with poorly activated nucleophilic partners, are clearly presented and projected in the context of asymmetric oganocatalysis. While many of these processes involve enamine or imine intermediates, secondary chiral amines were particularly studied as chiral organocatalyst in in this context; however the chapter also covers alternative organocatalysts acting as Brønsted acids or involving Hydrogen-Bonding such as cyclic amides, phosphoric acids or phosphonium salt.

Chapter 2 focuses on a specific type of organocatalyst, namely chiral quaternary ammonium salts ($R_4N^+X^-$). By acting as phase transfer catalyst, they make it possible to perform various transformations. Being inherently charged, they are able to form chiral ion-pairs or to act as Brønsted or Lewis bases. A detailed overview of the use of chiral quaternary ammonium salts in organocatalysis, including mechanistic aspects and scope regarding the type of created bond, will be provided in this chapter.

It is hoped and strongly believed that these chapters will provide the reader with valuable information on two aspects of a topic of major significance in modern organic chemistry, and will certainly stimulate its use as well as new developments in the future.

Luc NEUVILLE
Géraldine MASSON
April 2017

SOMO and Photoredox Asymmetric Organocatalysis

1.1. Introduction

Asymmetric organocatalysis, together with metal catalysis and biocatalysis, by responding to the concept of *Green Chemistry*, aims to overcome to the sustainability challenge which our modern society must face over in the forthcoming decades. Accordingly, this particularly important research topic has flourished over the past 15 years, as witnessed by the numerous achievements attained so far. From its renaissance at the beginning of the century, this field of research has constantly evolved. While initially dominated by enamine-based processes, the developments of iminium, Brønsted acid, *N*-heterocyclic carbene (NHC), ion-pairing, and numerous other strategies have allowed the boundaries of this catalytic method to progressively push forward and, thereby, step by step, have made it possible to solve many synthetic bottlenecks. Accordingly, the success of organocatalysis partly comes from its multifaceted nature that has allowed the advent of many different, yet complementary, modes of activation. Lately, the organic synthetic community could not help but notice a number of substantial advances in this field, in which, in opposition to commonly employed polar modes of activations organocatalysis has been merged with radical chemistry. The aim of this chapter is to highlight some of these innovative developments and hopefully demonstrate that, by harnessing the high reactivity of radical species, organocatalysis will definitely play a key role in the future. This review, which will primarily focus on asymmetric

Chapter written by Maxime R. VITALE.

transformations, will cover two major radical organocatalytic methods, namely, singly occupied molecular orbital (SOMO) and photoredox organocatalysis.

1.2. Asymmetric SOMO organocatalysis

The concept of SOMO activation was first introduced by MacMillan and co-workers in 2007 and has since been applied in a significant number of elegant enantioselective α-functionalization of aldehydes (or ketones) as well as cascade transformations. According to this concept, bond formation occurs by means of radical coupling with various nucleophilic partners and operates with an overall "*Umpolung*" (reversal of polarity) process. Before reviewing the synthetic potential of this organocatalytic mode of activation, we will first present the originality of this new concept.

1.2.1. *The concept of SOMO organocatalysis*

For several years, amine-based organocatalysis has been restricted to polar modes of activation in which carbonyl compounds are transiently transformed into enamine or iminium species. On the one hand, the generation of an enamine from an aldehyde (or a ketone) allows its nucleophilic character to be enhanced and encourages its interaction with electrophiles in position α (highest occupied molecular orbital [HOMO] activation, Scheme 1.1(a)). On the other hand, a catalytically generated α,β-iminium possesses a less energetic lowest unoccupied molecular orbital susceptible, in position β, to better interact with nucleophiles (lowest unoccupied molecular orbital [LUMO] activation, Scheme 1.1(b)). Despite the fact that these two generic modes of activation have been the key to the success of many useful synthetic transformations, the lack of reactivity of enamines/iminiums toward relatively nonpolar partners had long remained noticeably troublesome, so that another paradigm needed to be found.

This is in this particular context that, in 2007, MacMillan's and Sibi's group independently presented a third aminocatalytic activation mode [BEE 07, SIB 07]. Baptized "SOMO organocatalysis" by MacMillan and co-workers, this original concept relies on the idea that, upon the selective monoelectronic oxidation of a transient enamine, the generation of a 3π-electron radical cation should widen the scope of aldehyde α-functionalization. Indeed, as it had been already reported in the literature

[NAR 92, COS 93a, COS 93b], such species that possess a SOMO activation (Scheme 1.1(c)) are inherently very reactive electrophilic intermediates inclined to react via radical pathways with poorly activated nucleophilic partners.

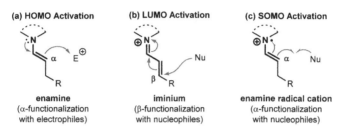

(a) HOMO Activation

enamine
(α-functionalization
with electrophiles)

(b) LUMO Activation

iminium
(β-functionalization
with nucleophiles)

(c) SOMO Activation

enamine radical cation
(α-functionalization
with nucleophiles)

Scheme 1.1. *Amine-based activations of carbonyl compounds*

The success of this new catalytic strategy was a real tour de force as it was conditioned by the ability to (1) find suitable reaction conditions susceptible to selectively oxidize the catalytically generated enamine, (2) find organocatalysts susceptible to induce satisfactory levels of enantioselectivity and (3) find adequate reaction partners susceptible to both trap the transient enamine radical cations and subsequently ensure the regeneration of the organocatalyst.

1.2.1.1. *Choice of the oxidant*

In the course of their studies, MacMillan and co-workers demonstrated the likeliness of the required selective monoelectronic oxidation of enamines by comparing the ionization potentials (IPs) of an aldehyde (butanal), a secondary amine (pyrrolidine) and that of the resultant enamine (Scheme 1.2).

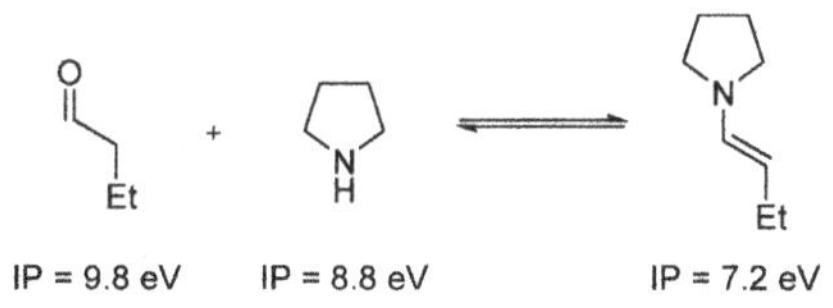

IP = 9.8 eV   IP = 8.8 eV   IP = 7.2 eV

Scheme 1.2. *Comparison of ionization potentials*

Although both the aldehyde and the amine (organocatalyst) possess high IP (>8 eV), it appears that the enamine, which possesses the highest HOMO, can reasonably be expected to be oxidized first (IP = 7.2 eV). Nevertheless, a careful choice of the oxidant is still required since the enamine is only catalytically generated in the reaction mixture and too strong an oxidant could concomitantly consume the starting materials and/or the organocatalyst. To date, cerium ammonium nitrate (CAN) has typically been used as oxidant in such SOMO organocatalytic processes, while in some cases Cu(II) and Fe(III) salts have also been reported (*vide infra*). Worthy of note, CAN is a very strong oxidant but its sparing solubility in common organic reaction media, which often entails vigorous stirring and additional water, was revealed to account for its selective oxidative action [DEV 10].

1.2.1.2. *Choice of organocatalyst*

Another key point concerns the choice of the organocatalyst. Based on their previous studies concerning enamine organocatalysis, MacMillan and co-workers proposed that the use of second-generation MacMillan imidazolidinone organocalysts would be particularly adequate for such SOMO-based transformations. The enamine radical cation, which exists under two resonance forms, keeps an sp^2 character such that the use of imidazolidinones were postulated to efficiently discriminate the two pseudo-enantiotopic faces of the radical. Indeed, while the bigger imidazolidinone substituent generally lies next to the enamine α-CH bond so as to minimize allylic $A^{1,3}$ strains, the β-CH bond is typically oriented toward the smaller imidazolidinone substituent in order to minimize $A^{1,4}$ interactions. Accordingly, the sp^2 enamine radical cation preferentially adopts a conformation in which the smaller group shields one face of the 3π system so that the trapping of the radical by the SOMOphile favorably happens on the less hindered face (Scheme 1.3). This strategy has so far led to the α-functionalization of carbonyl compounds with particularly high enantiocontrol (*vide infra*).

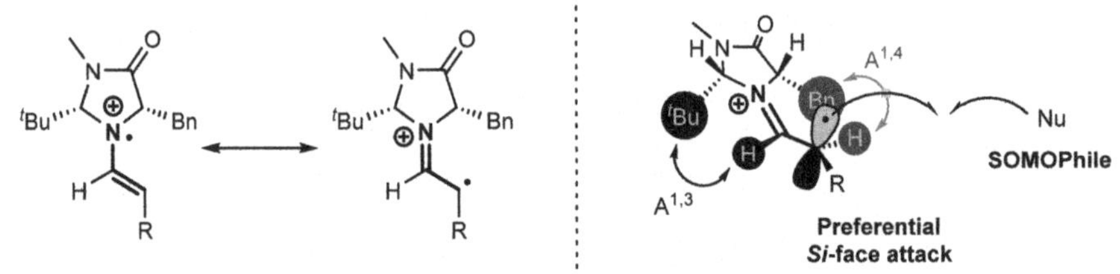

Scheme 1.3. *Stereoinduction by the organocatalyst*

1.2.1.3. *Choice of the reaction partner*

Last but not least, the success of a SOMO-organocatalytic transformation is also dependent on the nature of the reaction partner which is used. Indeed, the enamine radical cation possesses an electrophilic character due to the positive charge, which is created on the nitrogen atom upon oxidation. Accordingly, a suitable reaction partner (SOMOphile) should match this criterion and intrinsically possess a nucleophilic character whatever its polar or radical nature.

Additionally, after bond formation with a polar nucleophile, the resulting radical needs to be intercepted. While cascade reactions have been designed in which the transient radical is subsequently trapped by another polar partner (*vide infra*), the closing of the catalytic cycle generally entails an additional one-electron oxidation step such that an overall "umpolung" process takes place. Thereby, two equivalents of one-electron oxidant are usually required, unless the SOMOphile that is used is a radical itself (see section 1.2.2.8).

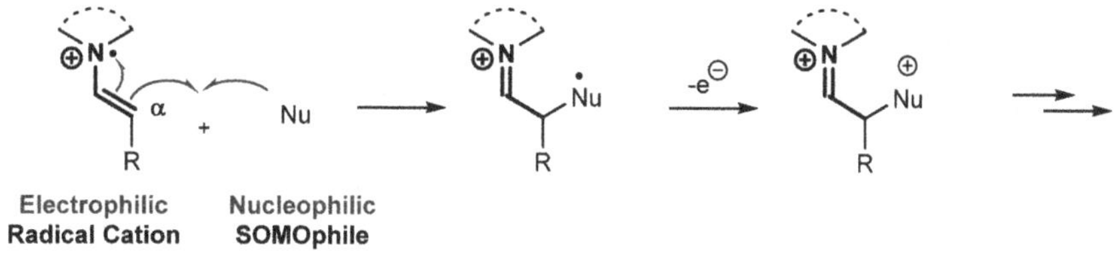

Electrophilic   **Nucleophilic**
Radical Cation   **SOMOphile**

Scheme 1.4. *Reaction between the enamine radical cation and the SOMOphile*

1.2.1.4. *General catalytic cycle*

With all these considerations, it becomes obvious that SOMO-organocatalysis is far from being trivial and that the success of such a catalytic strategy depends on many requirements and several elementary steps. Nevertheless, it is possible to consider a general catalytic cycle that fits most of the SOMO-organocatalytic transformations reported to date. For the sake of clarity, our description will be based on the α-allylation of aldehydes with allylsilanes mediated by CAN, a seminal example reported by MacMillan in 2007, which we will discuss later in more detail (Scheme 1.5) [BEE 07].

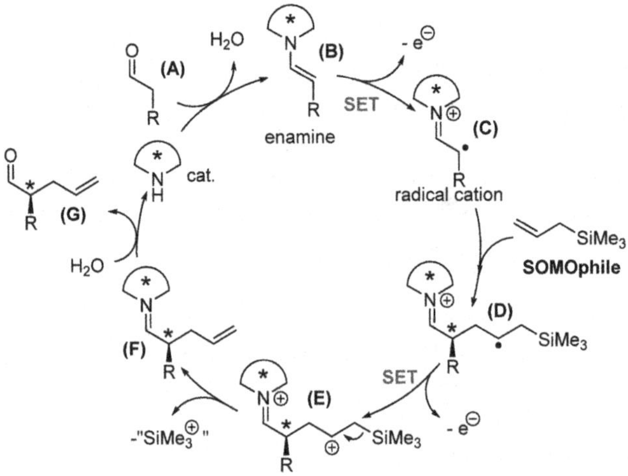

Scheme 1.5. *Typical SOMO-organocatalytic cycle*

At the onset of most SOMO-organocatalytic reactions, the aldehyde **A** and the secondary amine organocatalyst generate enamine **B**, which is engaged in a one-electron oxidation step to provide **C** [BEE 11]. This electrophilic radical then couples with a SOMOphile (allyltrimethylsilane in this case) to provide alkyl radical **D** which, after a second one-electron oxidation event, leads to the silicon-stabilized carbocation **E**. After the β-elimination of the silicon group, the resulting iminium **F** is hydrolyzed, thus allowing the formation of the desired α-allylated product **G** and the regeneration of the organocatalyst. Worthy of note, while this catalytic cycle is very general, it is not entirely consistent with all the transformations that we will discuss hereafter, which will lead us to give more details later on, whenever it is necessary.

1.2.2. *Enantioselective α-functionalization of carbonyl compounds*

1.2.2.1. *α-Allylations*

As discussed previously, one of the first SOMO-organocatalytic reactions reported in 2007 concerns the enantioselective α-allylation of aldehydes with functionalized allylsilanes [BEE 07]. In this seminal work, MacMillan and co-workers were able to demonstrate the feasibility of the SOMO-activation concept by submitting various aliphatic aldehydes to several allylsilanes in the presence of CAN (2 equiv.) in dimethoxymethane at −20 °C. It was found that the desired α-allylated products were obtained in good yields and excellent enantioselectivities when 20 mol% of the trifluoroacetic acid (TFA) salt of the second-generation MacMillan organocatalyst **O1** was used (Scheme 1.6).

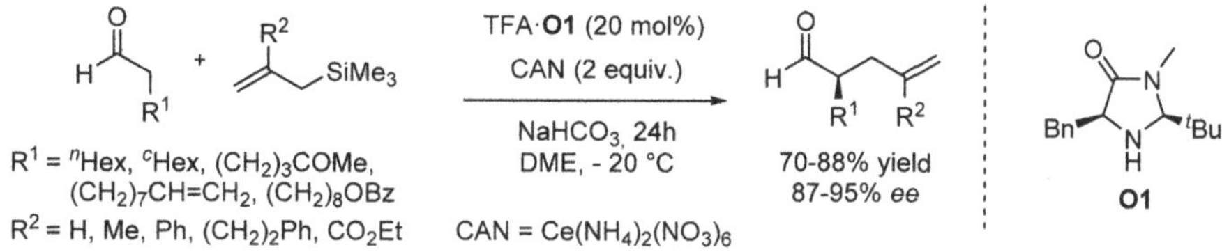

Scheme 1.6. *Intermolecular α-allylation of aldehydes with allylsilanes*

MacMillan and co-workers also showed that this organo-SOMO-catalytic α-allylation process is amenable to the use of ketones [MAS 10]. However, the development of specific imidazolidinones organocatalysts was required in this case. Indeed, while initial results employing (2*R*,5*R*)-5-benzyl-3-methyl-2-(5-methylfuran-2-yl)-imidazolidin-4-one (not shown) consistently gave poor yields of the desired allylated products due to the competitive oxidation of its furan ring by CAN, better results were obtained when employing the less electron-rich organocatalysts **O2–O4**. Accordingly, the α-allylated ketones were reliably obtained in 65–86% yield and 83–99% *ee*. Worthy of note, starting from unsymmetrical ketones ($n = 0$, $X \neq CH_2$) allylation regioselectively occurred at the C4-position and the use of cyclopentanone exclusively led to the corresponding *trans*-2,5-*bis*-allylated ketone (Scheme 1.7).

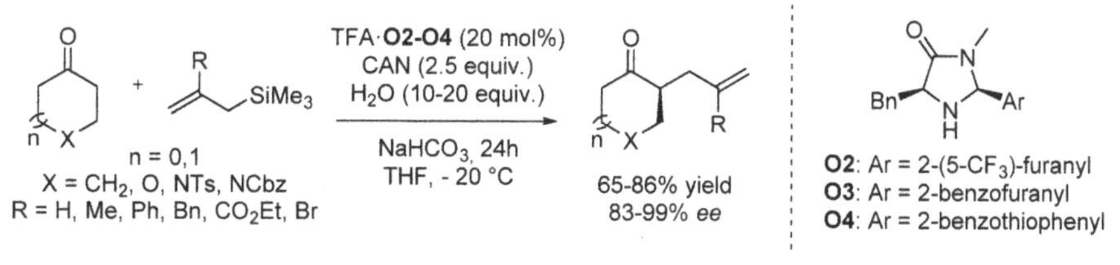

Scheme 1.7. *Intermolecular α-allylation of ketones with allylsilanes*

Later, the same group showed that the related intramolecular allylation reactions could be realized starting from allylsilane-tethered aldehydes, CAN

(2.2 equiv.) and TFA·**O1** (20 mol%) [PHA 11]. Interestingly, the presence of water (2 equiv.) had a salient effect on the yield, the diastereoselectivity and the enantioselectivity of the targeted cyclization process. Although the authors suggested that this extra amount of water may help the desilylation process, it is also likely that it helps the regeneration of the organocatalyst and that it increases the global concentration of CAN in the reaction mixture [DEV 10]. In some cases, better results were obtained when using acetone as solvent and 2,6-di-*tert*-butylpyridine (DTBP) as base. The corresponding carbo- and heterocycles were typically obtained in good yields, good enantiopurities, and high diastereoselectivities in favor of the *trans* diastereoisomers (Scheme 1.8).

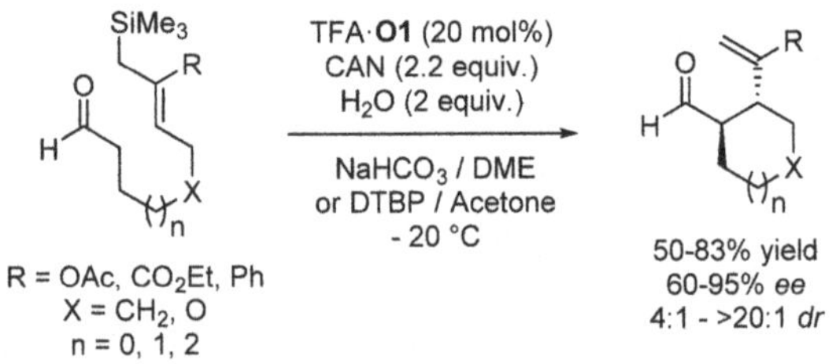

Scheme 1.8. *Intramolecular α-allylation of aldehydes with allylsilanes*

More recently, MacMillan and co-workers successfully extended this intramolecular α-allylation of aldehydes to the use of simple alkene-tethered aldehydes [COM 13]. In this case, the optimization studies revealed that the bulkier 1-naphthyl organocatalyst **O5** ensured the best levels of enantioselectivity and that better yields were typically obtained when the potent 1-electron oxidant and very much soluble *tris*-bistriflimide salt of Fe(III) *tris*-phenanthroline was used. Considering that it is necessary to transiently generate a carbocation, it is worth noting that this method remains limited to the use of trisubstituted alkenes. Nevertheless, it allowed a good range of carbo- and heterocyclic systems in good to excellent yields and enantioselectivities to be prepared and again, as a result of chair-like transition states, excellent *trans* diastereocontrols were typically obtained (Scheme 1.9).

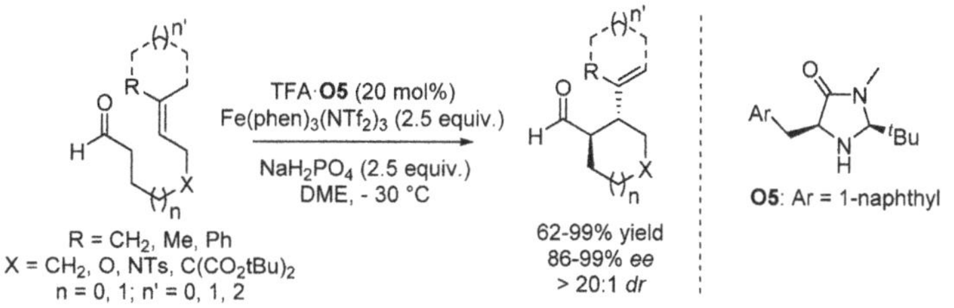

Scheme 1.9. *Intramolecular α-allylation of aldehydes with alkenes*

1.2.2.2. α-Enolations

In 2007, shortly after showcasing the SOMO-organocatalyzed α-allylation of aldehydes, MacMillan and co-workers reported that, under similar reaction conditions, the use of enol silanes, the oxygenated analogues of allylsilanes, readily opened the way to the enantioselective α-enolation of aldehydes (Scheme 1.10) [JAN 07].

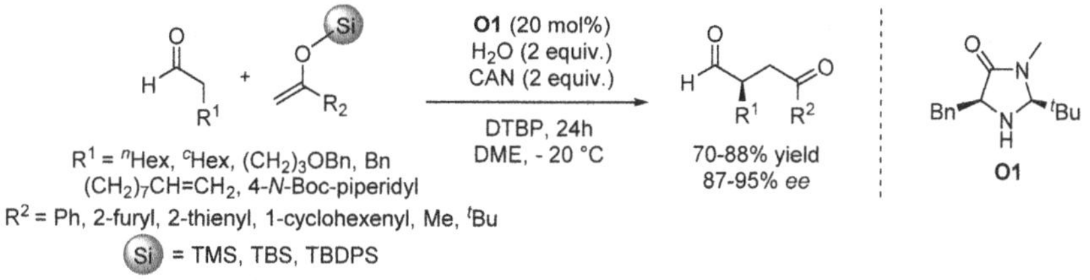

Scheme 1.10. *Intermolecular α-enolation of aldehydes*

Excellent enantioselectivities could be attained when **O1** was used as organocatalyst, while the global efficiency of the reaction was strongly enhanced by the presence of 2 equiv. of water and DTBP.

From a mechanistic point of view, in parallel to what was proposed by Murakami and co-workers [NAR 92], the transient enamine is first engaged in a one-electron redox process with CAN and thereby generates the corresponding enamine radical cation. This electrophilic species then preferentially couples with the enolsilane from its *Si* face so as to minimize steric interactions with the imidazolidinone substituents. The resulting α-silyloxy radical subsequently undergoes a second one-electron oxidation event and, after desilylation and hydrolysis, the α-enolated aldehyde is released together with regeneration of the organocatalyst (Scheme 1.11).

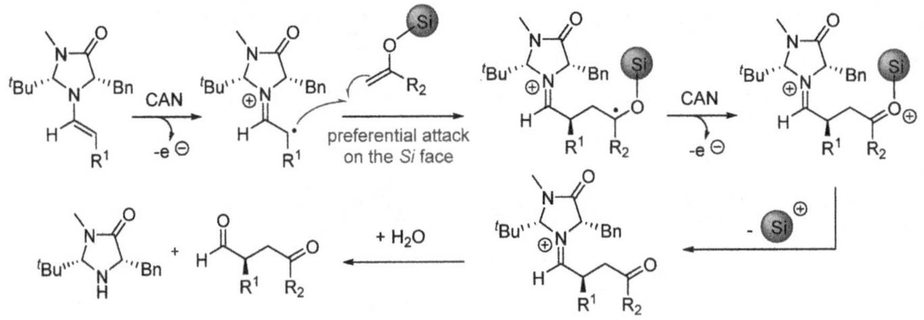

Scheme 1.11. *Mechanistic aspects of the α-enolation of aldehydes*

The α-enolation of ketones has, in comparison, been scarcely reported. MacMillan and co-workers described a single example based on the use of cyclohexanone and organocatalyst **O4** (73% yield, 84% *ee*) [MAS 10], while recent attempts employing cyclopentanone and cyclohexanone with other imidazolidinone catalysts resulted in negligible diastereo- and enantioselectivities [TIS 14]. Alternatively, the use of silyl ketene acetals and silyl ketene thioacetals was explored by Šebesta and co-workers and usually gave moderate enantioselectivities and yields [TIS 14].

1.2.2.3. α-Homobenzylations with styrenes

In 2008, MacMillan and co-workers showed that styrenes may also participate as SOMOphiles and may allow, thereby, the formal enantioselective α-alkylation of aldehydes [GRA 08]. Under typical TFA•**O1**/CAN/H$_2$O SOMO-organocatalytic conditions, the desired homobenzylated aldehydes were obtained as diastereomeric mixtures of benzylic nitro esters (approximately 3:1 *dr*), good yields (up to 95%) and excellent enantioselectivities (up to 95%) (Scheme 1.12).

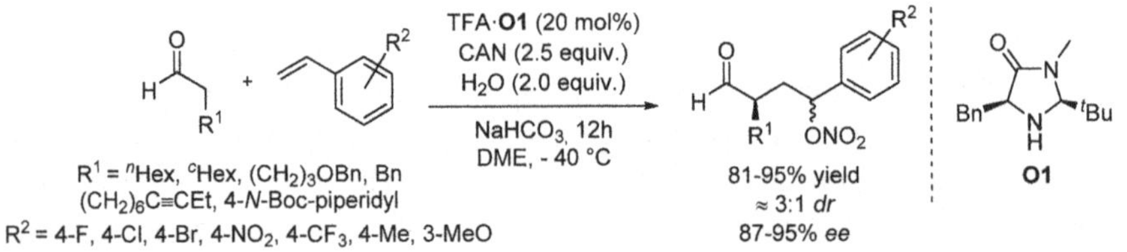

Scheme 1.12. *α-Homobenzylations of aldehydes with styrenes*

From a mechanistic point of view, the authors propose that, after radical coupling of the enamine radical cation with the styrene partner, a further one-electron oxidation event transiently generates a benzylic carbocation, which is subsequently trapped by a nitrate anion coming from the oxidant (CAN) (Scheme 1.13).

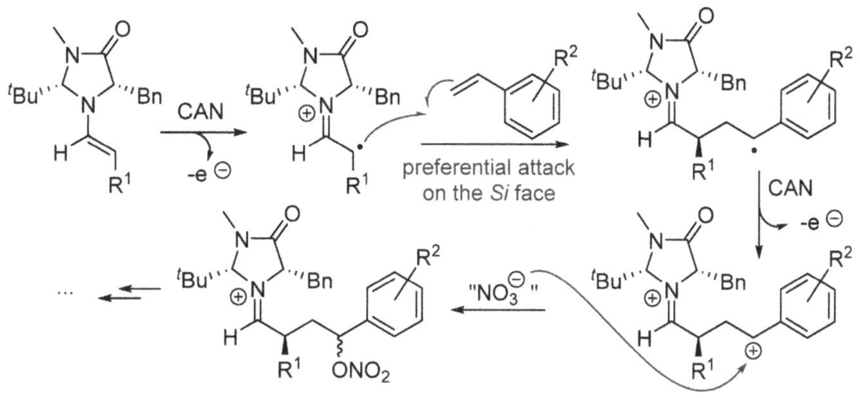

Scheme 1.13. *Mechanistic proposal for the α-homobenzylations with styrenes*

Interestingly, while the corresponding benzylic nitro esters were obtained with moderate diastereoselectivities, the authors disclosed a two-step procedure in which the initial SOMO-organocatalytic transformation is followed by a catalytic hydrogenation. This method allowed to the resulting α-homobenzylated aldehydes in high optical purity and good global yields be straight forwardly obtained, as a result of the succeeding hydrogenolitic cleavage of the nitrate ester C–O bond (Scheme 1.14).

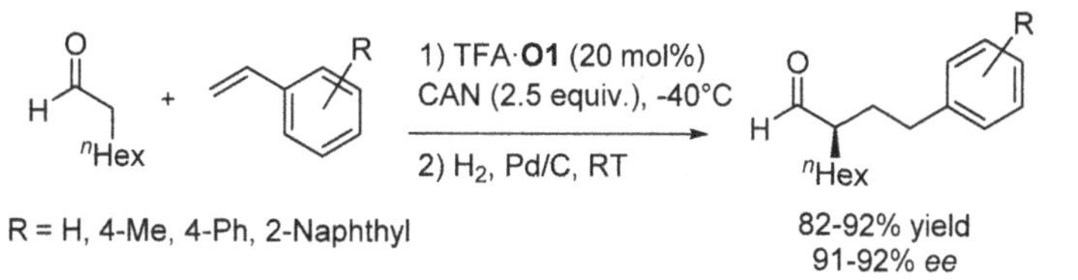

Scheme 1.14. *Two-step procedure for the α-homobenzylations of aldehydes*

1.2.2.4. α-Nitroalkylations with silylnitronates

In line with the previously successful α-functionalization of aldehydes with allyl- and enol-silanes, MacMillan and co-workers reported the use of silylnitronates which, in close oxidative reaction conditions, allowed the corresponding α-nitroalkylated aldehydes in good yields, good to excellent enantioselectivities and variable diastereoselectivities to be obtained (Scheme 1.15) [WIL 09].

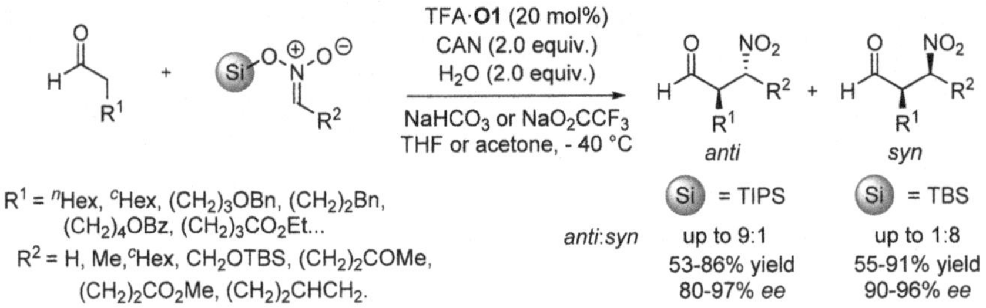
TFA·O1 (20 mol%)
CAN (2.0 equiv.)
H₂O (2.0 equiv.)
NaHCO₃ or NaO₂CCF₃
THF or acetone, - 40 °C
anti
syn
R¹ = ⁿHex, ᶜHex, (CH₂)₃OBn, (CH₂)₂Bn, (CH₂)₄OBz, (CH₂)₃CO₂Et...
R² = H, Me, ᶜHex, CH₂OTBS, (CH₂)₂COMe, (CH₂)₂CO₂Me, (CH₂)₂CHCH₂.
Si = TIPS
Si = TBS
anti:syn
up to 9:1
53-86% yield
80-97% ee
up to 1:8
55-91% yield
90-96% ee

Scheme 1.15. *α-Nitroalkylations of aldehydes with silylnitronates*

During the course of their study, the authors observed that the nitronate silyl protecting group had a direct influence on the diastereoselectivity of this transformation. Indeed, while *O*-TIPS (TIPS = tri*iso*propylsilyl) nitronates predominantly led to the *anti*-α-nitroalkylated aldehydes, the *O*-TBS analogs (TBS = *ter*butyldimethylsilyl) preferentially induced the formation of the related *syn* diastereoisomers. To account for this protecting group induced diastereoselectivity switch, the authors suggested that two different reaction mechanisms are involved, depending on the stability of the Si-O bond toward cleavage. On the one hand, the use of a robust *O*-TIPS nitronate was proposed to favor the desired organo-SOMO transformation through the direct *anti* selective coupling of the enamine radical cation with the nitronate (SOMO pathway, Scheme 1.15). On the other, a more labile *O*-TBS nitronate was suggested to primarily undergo desilylation and, upon the 1-electron oxidation of the resulting very electron-rich nitronate species, the corresponding electrophilic radical cation was thought to couple with the enamine (being in this case the SOMOphile) in an *syn* selective manner (SOMOphilic pathway, Scheme 1.16). However, this proposed mechanistic scenario does not clearly explain how each pathway influences the diastereoselectivity outcome of the corresponding coupling process.

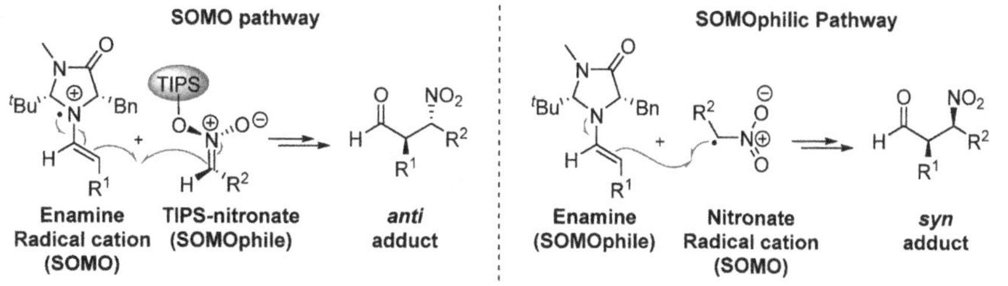
SOMO pathway
SOMOphilic Pathway
Enamine Radical cation (SOMO)
TIPS-nitronate (SOMOphile)
anti adduct
Enamine (SOMOphile)
Nitronate Radical cation (SOMO)
syn adduct

Scheme 1.16. *Mechanistic dichotomy of the α-nitroalkylations of aldehydes*

Hence, not only SOMO organocatalysis may be envisioned through the formation of enamine radical cations, but also via the radical couplings of catalytically generated enamine with the appropriate electrophilic radicals. This second strategy has been particularly developed under photoredox co-catalytic activation, a particular research topic that we will discuss later.

1.2.2.5. *α-Alkenylations*

SOMO-organocatalysis has also been central to the resolution of other longstanding bottlenecks in asymmetric synthesis such as the enantioselective α-alkenylation of aldehydes. In 2008, MacMillan's group reported the first examples of such transformation in which the organocatalytic activations of aldehydes in the presence of alkenyltrifuoroborate salts and CAN yielded the desired α-alkenylated products in excellent yields and enantioselectivities (Scheme 1.17) [KIM 08].

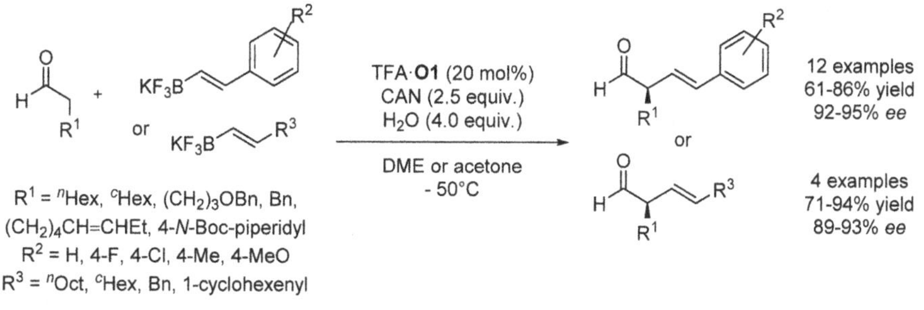
TFA·O1 (20 mol%)
CAN (2.5 equiv.)
H2O (4.0 equiv.)
DME or acetone
- 50°C
12 examples
61-86% yield
92-95% ee
4 examples
71-94% yield
89-93% ee
R1 = nHex, cHex, (CH2)3OBn, Bn,
(CH2)4CH=CHEt, 4-N-Boc-piperidyl
R2 = H, 4-F, 4-Cl, 4-Me, 4-MeO
R3 = nOct, cHex, Bn, 1-cyclohexenyl

Scheme 1.17. *α-Alkenylations of aldehydes with potassium alkenyltrifluoroborates*

While this study mainly concerned the use of styryl potassium trifluoroborates, it was shown that some alkyl-, cyclohexyl- and

1-cyclohexyl-substituted counterparts also performed well in this original coupling reaction. In all cases, the α-alkenylated aldehydes were obtained as perfectly pure (*E*) isomers, which the authors rationalized by a *trans* selective Peterson-like elimination consecutive to C–C bond formation and 1-electron oxidation of the resulting β-boryl radical (Scheme 1.18).

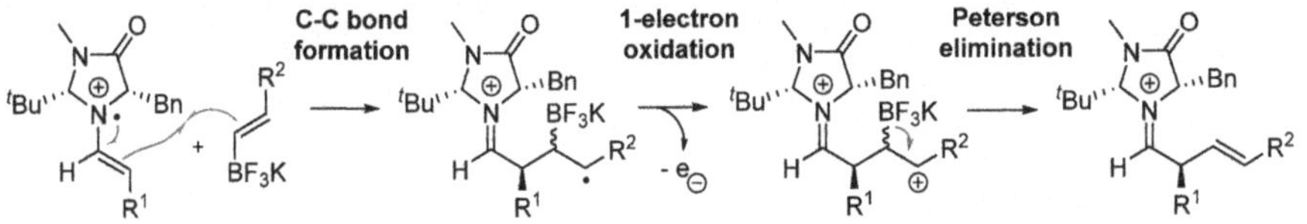

Scheme 1.18. *Mechanistic proposal for the α-alkenylations of aldehydes*

1.2.2.6. α-Arylations

While in their very first report about SOMO-organocatalysis MacMillan and co-workers described the successful intermolecular coupling of *N*-Boc pyrrole with octanal [BEE 07], in 2009, both MacMillan's and Nicolaou's group almost simultaneously reported the intramolecular α-arylations of aldehydes [NIC 09, CON 09]. In very close reaction conditions, these research teams demonstrated that various δ-aryl- or δ-heteroaryl-substituted aldehydes underwent an intramolecular pseudo-dehydro-coupling reaction leading to the corresponding α-arylated products. MacMillan and co-workers systematically observed better enantioselectivities when the 1-naphthyl organocatalyst **O5** was used. However, the more readily available imidazolidinone **O1** still induced very good enantiocontrol (Scheme 1.19).

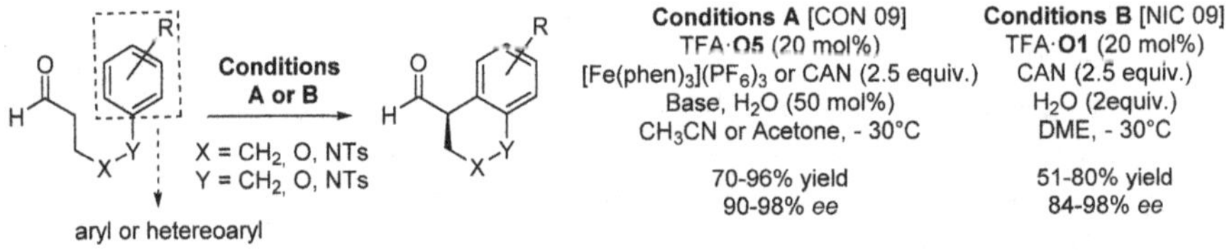

Scheme 1.19. *Intramolecular α-arylations of aldehydes*

From a mechanistic perspective, it was initially proposed by Nicolaou *et al.* that this transformation occurred through an intramolecular Friedel–Crafts reaction triggered by the formation of a carbocation resulting from the oxidation of the transient enamine. However, after careful density functional theory studies, MacMillan *and co-workers* were able to establish that it belonged to the family of SOMO-organocatalyzed processes, in which a 3π-electron radical cation interacts with the aromatic moiety according to a radical-based attack (Scheme 1.20) [UM 10].

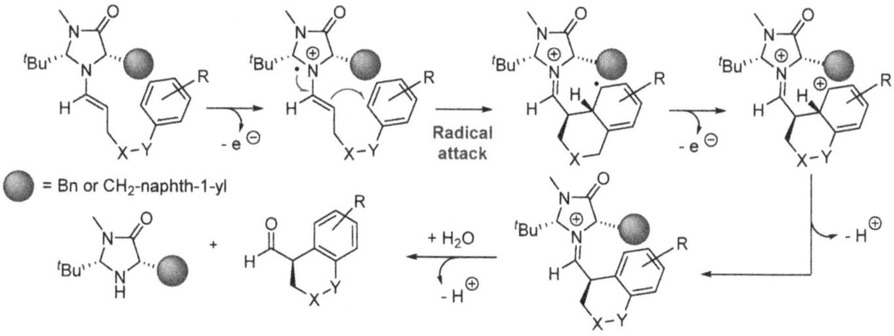

Scheme 1.20. *Mechanistic proposal for the intramolecular α-arylations of aldehydes*

The efficiency of this method was recently highlighted by Li and co-workers in the context of the total synthesis of pseudo-pteroxazole, for which one of the key steps is such a SOMO-organocatalyzed α-arylation reaction (Scheme 1.21) [YAN 16].

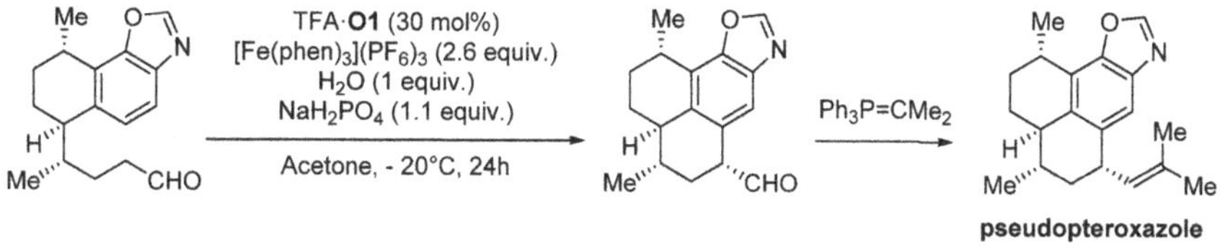

Scheme 1.21. *Application of the α-arylation of aldehydes in total synthesis*

1.2.2.7. α-Chlorinations

In 2009, as a continuation of their interest in the enantioselective α-chlorination of aldehydes [BRO 04], MacMillan and co-workers reported that this kind of transformation could be performed, under oxidative organocatalysis, without the need for electrophilic sources of chlorine [AMA 09]. Under imidazolidinone organocatalysis and employing cheap lithium chloride in the presence of stoichiometric amounts of a single electron oxidant, aldehydes were shown to straightforwardly afford the corresponding α-chloro aldehydes. While initial results based on the use of organocatalyst **O1** and CAN afforded good yield and good enantioselectivity at low temperature (−78°C), further studies showed that it was not suitable when the reaction was performed at room temperature since it progressively led to the erosion of the product enantiopurity over the course of the reaction. For this reason, based on density functional theory (DFT) studies, the authors designed organocatalyst **O6**, with which the enantioselective α-chlorination of aldehydes occurred at 10°C, without concurrent epimerization, when employing a LiCl/Cu(TFA)$_2$/ Na$_2$S$_2$O$_8$ system. Under these conditions, the targeted α-chlorinated aldehydes were typically obtained in good yields and excellent enantioselectivities (Scheme 1.22).

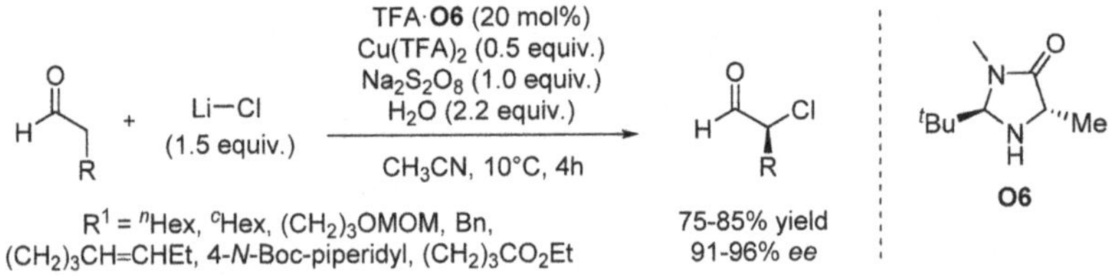

TFA·**O6** (20 mol%)
Cu(TFA)$_2$ (0.5 equiv.)
Na$_2$S$_2$O$_8$ (1.0 equiv.)
H$_2$O (2.2 equiv.)

Li—Cl
(1.5 equiv.)

CH$_3$CN, 10°C, 4h

R^1 = ^{n}Hex, ^{c}Hex, (CH$_2$)$_3$OMOM, Bn,
(CH$_2$)$_3$CH=CHEt, 4-N-Boc-piperidyl, (CH$_2$)$_3$CO$_2$Et

75-85% yield
91-96% ee

Scheme 1.22. *Oxidative α-chlorination of aldehydes with lithium chloride*

Worthy of note, although the authors suggested that a SOMO-organocatalyzed process takes place, the required use of both sodium persulfate and copper(II) trifluoroacetate was not perfectly clarified. Accordingly, a polar reaction mechanism in which *in situ* generated copper(II) chloride acts as an electrophilic chlorine source cannot be fully ruled out. Indeed, Owsley and co-workers previously demonstrated that CuCl$_2$ allowed the racemic α-chlorination of aldehydes [CAS 65]. If so, sodium persulfate could serve the regeneration of CuCl$_2$ by oxidation of CuCl in the presence of LiCl.

1.2.2.8. α-Oxyaminations

About the same time as MacMillan and co-workers first communicated on SOMO-organocatalysis [BEE 07], Sibi and Hasegawa reported the enantioselective α-oxyamination of aldehydes by employing oxidative organocatalytic conditions [SIB 07]. Initial reaction conditions based on the catalytic use of the imidazolidinone organocatalyst **O7** and a stoichiometric amount of ferrocenium tetrafluoroborate in the presence of TEMPO (2,2,6,6-tetramethylpiperidine *N*-oxide) were allowed to perform, at room temperature in THF, the α-oxyamination of hydrocinnamaldehyde in 87% yield and 80%

ee. Upon further optimization, these authors showed that catalytic quantities of $FeCl_3$ and $NaNO_2$ under oxygen atmosphere in DMF efficiently replaced the ferrocenium salt and that the resulting catalytic system permitted the α-oxyaminations of a good range of aldehydes with average to excellent yields and enantioselectivities (Scheme 1.23).

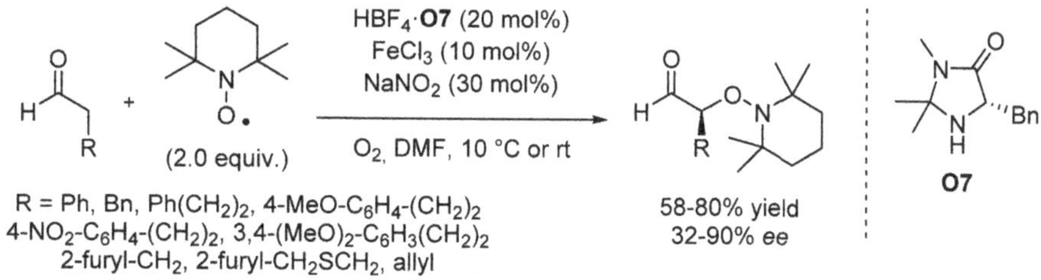
HBF4·O7 (20 mol%)
FeCl3 (10 mol%)
NaNO2 (30 mol%)
O2, DMF, 10 °C or rt
(2.0 equiv.)
R = Ph, Bn, Ph(CH2)2, 4-MeO-C6H4-(CH2)2
4-NO2-C6H4-(CH2)2, 3,4-(MeO)2-C6H3(CH2)2
2-furyl-CH2, 2-furyl-CH2SCH2, allyl
58-80% yield
32-90% ee
O7

Scheme 1.23. *$FeCl_3$-mediated α-oxyamination of aldehydes*

Sibi and Hasegawa proposed a reaction mechanism in which the enamine, formed by the condensation of the organocatalyst **O7** with the aldehyde partner, undergoes a single electron oxidation by $FeCl_3$ to generate a 3π-electron radical that, in a radical–radical coupling manner, is subsequently trapped by TEMPO. These authors also suggested that this transformation could efficiently be realized with catalytic quantities of $FeCl_3$ since $NaNO_2$ allowed the back oxidation of Fe(II) to Fe(III) under oxygen atmosphere (Scheme 1.24).

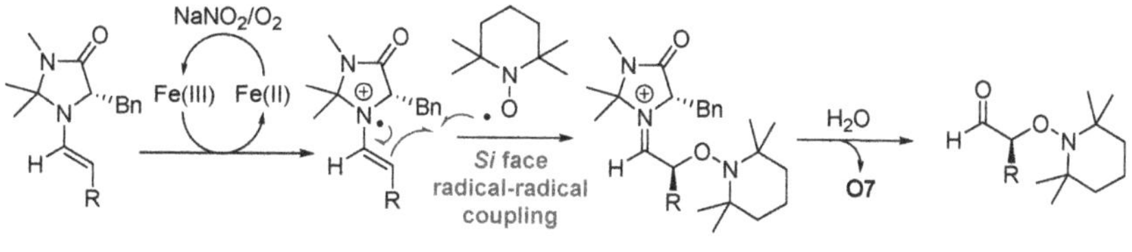
NaNO2/O2
Fe(III) Fe(II)
Si face
radical-radical
coupling
H2O
O7

Scheme 1.24. *Sibi's mechanistic proposal for the Fe(III)-mediated α-oxyamination of aldehydes*

While this mechanistic picture is perfectly in line with the principle of SOMO-organocatalysis, it was later shown by MacMillan and co-workers that, in fact, a non-radical polar reaction mechanism most probably operates in this $FeCl_3$-mediated α-oxyamination reaction [VAN 10]. Indeed, although this reaction entailed the use of DMF, it is well established that this particular solvent induces disproportionation of $FeCl_3$ to $[Fe(DMF)_3Cl_2]$ $[FeCl_4]$ and, by means of cyclic voltammetry studies, it was demonstrated that the oxidation potential of the latter does not thermodynamically allow the required oxidation of the enamine [TOB 72]. However, with regard to Sibi's ferrocenium tetrafluoroborate mediated α-oxyamination of hydrocinnam-aldehyde, no substantial evidence permits a SOMO-organocatalytic type process to be completely ruled out and, for this reason, the paternity of SOMO-organocatalysis could arguably be shared between MacMillan and Sibi.

Based on their mechanistic studies, MacMillan and co-workers later reported a competitive metal/organo synergistic catalytic system relying on the aerobic use of $CuCl_2$ and an indole-derived imidazolidinone [SIM 12].

1.2.3. *Enantioselective cascade and pseudo-cycloaddition processes*

SOMO-organocatalysis has also undergone some exciting developments in the context of cascade and pseudo-cycloaddition processes. As opposed to the previously described α-functionalization reactions of carbonyl compounds, which mostly concern the radical creation of one new chemical bond, MacMillan and co-workers showcased the possibility of harnessing the high reactivity of the resulting transient radicals (or carbocations), so as to promote multiple bond forming processes.

1.2.3.1. *Polyene cyclizations*

Polyene cyclizations constitute one of the major biosynthetic pathways of naturally occurring terpenes and, accordingly, have been the focus of much research [YOD 05]. In addition to the plethora of biomimetic cationic transformations that have been developed, the akin radical-based processes have also known significant endeavors [DHI 01]. It is in this particular context that, in 2010, MacMillan and co-workers reported the first highly diastereo- and enantioselective SOMO-organocatalyzed polyene cyclizations [REN 10].

Employing polyunsaturated aldehydes and imidazolidinone **O1** with either CAN or $[Fe(phen)_3][PF_6]$ did not allow the desired cascade reaction to take place. Indeed, in these cases, products derived from premature oxidation of the tertiary alkyl radical intermediate or capture of the radical by external nucleophiles were predominant. However, upon the screening of other oxidant sources, it was found that the $Cu(OTf)_2$/NaTFA/TFA system, when slowly added to the reaction mixture, permitted the desired enantioselective cascade cyclization with organocatalyst **O5**. In the corresponding reaction conditions, several bicyclization products could be obtained with satisfactory yields and good levels of enantioselectivity (Scheme 1.25).

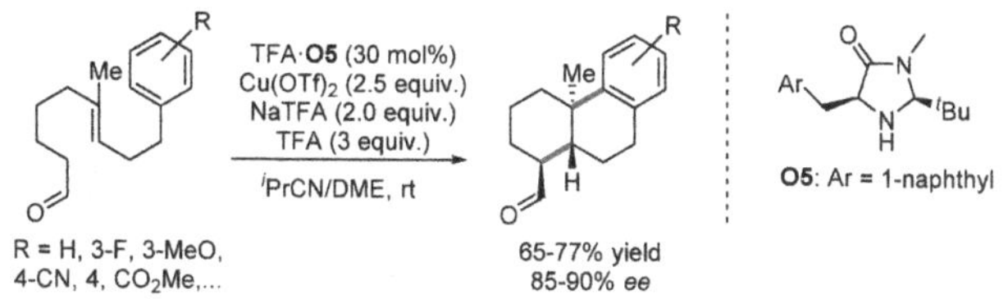

Scheme 1.25. *Cascade bicyclizations under SOMO-organocatalysis*

As a direct continuation of this work, the authors evaluated this SOMO-organocatalytic method for more complex polycyclization reactions. To this aim, extended polyene substrates were rationally designed as it was anticipated that the electronic properties of the tethered aldehyde would play a pivotal role in promoting the desired single-electron pathway toward cascade ring construction in lieu of unproductive early radical oxidation or non-regioselective alkene addition. Hence, polycycle precursors incorporating an alternating sequence of polarity-inverted C = C bonds (acrylonitrile and isobutene residues) were chosen in order to match the electrophilic/nucleophilic nature of the transient radical intermediates with the adequate nucleophilic/electrophilic character of the unsaturation to which they should couple with. This strategy met great success and allowed the enantioselective preparation of several tricyclization products as well as polycyclized ones (Scheme 1.26).

As a brilliant feat of this methodology, the synthesis of a heptacycle through the generation of six new C–C bonds, 11 contiguous stereogenic centers (including five all-carbon quaternary stereocenters) was realized in a single step with 62% yield (Scheme 1.26) [REN 10].

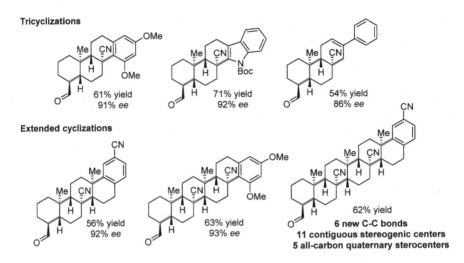

Scheme 1.26. *Tricyclizations and extended polycyclizations under SOMO-organocatalysis*

1.2.3.2. *Pseudo-cycloaddition processes*

MacMillan and co-workers formerly reported that the CAN-mediated SOMO-organocatalyzed coupling of aldehydes with styrenes led to the formation of the corresponding α-homobenzylated aldehydes nitro esters according to a radical-polar cross-over mechanism (see section 1.2.2.3). Based on this observation, the same group postulated that the fleeting benzylic cation should be vulnerable to an intramolecular nucleophilic addition so as to induce a cascade cycloaddition process.

Following this hypothesis, they envisioned that employing aldehydes possessing a pendent nucleophilic moiety would, according to a SOMO-organocatalytic pathway, lead to a 3π-radical cationic species susceptible to couple with a styrene partner so as to generate a benzylic radical intermediate. After a subsequent one-electron oxidation, whereby a highly electrophilic benzylic carbocation would be formed, the intramolecular attack of the nucleophilic residue was expected to trigger the desired ring closure and eventually lead, upon hydrolysis of the transitory iminium, to the targeted cycloadducts (Scheme 1.27).

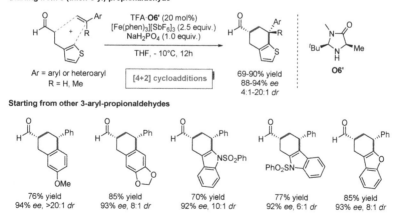

Scheme 1.27. *General approach to SOMO-organocatalyzed cascade cycloadditions*

At first, the SOMO-organocatalyzed [4+2] cycloadditions of 3-aryl-propionaldehydes with styrenes was investigated [JUI 10]. After a proper screening of different oxidant sources and organocatalysts, it was shown that the desired "olefin-addition/Friedel–Crafts" cascade could be achieved by using 2.5 equiv. of [Fe(phen)₃][SbF₆]₃ and **O6'**. In these reaction conditions, 3-(thien-3-yl)-propionaldehyde efficiently coupled with a wide range of styrene partners with good to excellent yields and stereoselectivities, while other electron-rich benzenes and heteroarenes (anisoles, catechols, indoles and benzofurans) were similarly demonstrated to be suitable nucleophilic terminators (Scheme 1.28).

Scheme 1.28. *SOMO-organocatalyzed [4+2] cycloadditions of 3-aryl-propionaldehydes*

While in the same report the authors demonstrated that, in two cases, allylsilane-tethered aldehydes similarly induced the desired cycloaddition process with styrenes [JUI 10], it was later shown that inner nitrogen-based nucleophiles opened the way to the diastereo- and enantioselective synthesis of pyrrolidine derivatives [JUI 12]. During the evaluation of various N-protected 3-amino aldehydes as formal dipoles, MacMillan and co-workers demonstrated that the N-nosyl derivative efficiently coupled with a wide range of styrenes according to a radical/polar cross-over [3+2] cycloaddition mechanism. In reaction conditions close to those used for 3-aryl-propionaldehydes, the corresponding pyrrolidines were consistently obtained with moderate to good yields and diastereoselectivities as well as excellent enantiocontrol (Scheme 1.29).

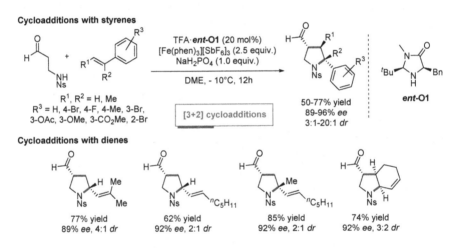

Scheme 1.29. SOMO-organocatalyzed [3+2] cycloadditions
of N-protected 3-amino-propionaldehydes

Worthy of note, this cycloaddition process could be applied to the use of heteroaromatic styrene derivatives and, more importantly, to the use of SOMOphilic dienes. In the second case, while good yields and enantioselectivities were attained, the diastereoselectivity of the overall [3+2] cycloaddition process was typically lower than that observed with styrenes.

1.2.4. *Conclusion*

Since its discovery in 2007, the SOMO-organocatalytic activation has witnessed numerous developments. Not only has it allowed the boundaries of amine-based organocatalysis to be pushed forward, but it has also revolutionized our perception of organocatalysis in a more general way. Indeed, this new paradigm in which organocatalysis is compatible with radical chemistry has served as a strong incentive for the development of powerful catalytic methods, such as asymmetric photoredox organocatalysis.

1.3. Asymmetric photoredox organocatalysis

As previously discussed, SOMO-activation has changed the perception of amine-based organocatalysis and has demonstrated the tremendous opportunities offered by harnessing the high reactivity of radicals in organocatalytic bond forming processes. Nevertheless, it is also true that, on sustainability grounds, the requirement of at least 2 equiv. of single electron oxidant may question the industrial application potential of this catalytic method. Although alternative monoelectronic oxidation methods based on the use of electrosynthesis have been explored, they have so far met limited success [BUI 09, HO 10]. Hence, the development of radical organocatalytic methods, in which an overall redox-neutral process can take place, has been the focus of important research.

It is in this particular context that the concept of combining photochemistry with organocatalysis was born. Based on the fact that light may transfer its energy to molecules and that the corresponding photoactivated compounds typically possess SOMO orbitals that may be prone to be engaged in single electron transfers (SETs), it was anticipated that it would be an efficient method to generate radicals in a very sustainable manner so as to use them in organocatalytic processes. Yet, because most organic compounds require high-energy wavelengths to reach excitation and that the corresponding excited states are usually short-lived, this particular field of research has mostly flourished through the use of an extra photoredox catalyst which, in addition to possessing a longer excited lifetime, can reach photoexcitation upon irradiation with visible light. Over the past 8 years, this synergistic catalytic method, often referred to as *"Photoredox Organocatalysis"*, has revolutionized the organocatalysis landscape.

After a basic introduction to the concept of photoredox catalysis, we will present its synergistic association with asymmetric organocatalysis as well as some alternative organocatalytic strategies mediated by light. This section, which does not aim to exhaustively review the progress in this area, will mainly focus on light-mediated asymmetric transformations in which the key bond formation step occurs according to a radical mechanism. Hence, cascade reactions in which photoredox catalysis is combined with polar organocatalysis will not be presented.

1.3.1. Photoredox catalysis

Over the last century, directly inspired by the emblematic use of *photosynthesis* by the vegetal kingdom, the concept of harnessing the energy of visible light has been the focus of a great amount of research. In the last four decades, several significant achievements have been attained in different chemistry areas that encompass water splitting, carbon dioxide reduction and photovoltaics [BAL 15].

In the late 1970s, Kellogg's and Deronzier's groups independently discovered that the catalytic use of a transition metal-based (or a purely organic) dye could efficiently allow, upon visible-light photon absorption, the development of unprecedented organic synthetic transformations [HED 78, BER 79, CAN 84a, CAN 84b]. By converting visible light into chemical energy and by promoting the generation of open-shell reactive species from organic substrates, it was demonstrated that *"photoredox catalysis"* could allow original reactivity patterns to be reached, which would not be conceivable otherwise. Regrettably, despite its wonderful potential for organic synthesis, this catalytic strategy has subsequently remained underappreciated for more than 30 years. However, because of its remarkable renaissance in the late 2000s, *"photoredox catalysis"* was, by the end of 2016, at the forefront of organic chemistry [ROM 16, SHA 16, KOI 14, PRI 13]. Importantly, rather than giving a full overview including all intricate photophysical details, our aim is to only present the basics of this principle, so as to ease the reader's comprehension of its synergistic association with organocatalysis.

A major breakthrough in the field of photocatalysis came from the identification of metal polypyridyl complexes as remarkably efficient photocatalysts (PCs). In particular, the corresponding ruthenium- and iridium-based complexes (such as $Ru(bpy)_3Cl_2$ and $Ir(ppy)_3$) revealed themselves particularly well as they, upon irradiation with visible light, undergo a metal to ligand charge transfer (CT) MLCT (one electron of the metal-centered HOMO jumps to an antibonding ligand-centered LUMO) followed by a non-radiative intersystem crossing ISC (spin polarity inversion) (Scheme 1.30).

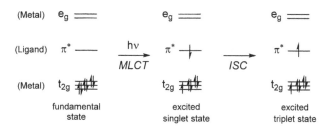

Scheme 1.30. *Simplified molecular orbital depiction of the photoexcitation process of metal polypyridyl complexes*

One great advantage of such photocatalyst (PC) is that, after intersystem crossing, its radiative relaxation is slowed down due to the spin-forbidden character of the phosphorescence process. Accordingly, its triplet photoactivated state typically exhibits a relatively long lifetime, a specificity that offers the opportunity to exploit its redox properties toward organic molecules. Indeed, according to an outer-sphere mechanism, the metastable photocatalyst (^{3}PC*) becomes able to either give or take an electron (oxidative/reductive quench, respectively) to or from an appropriate substrate (provided a match of the redox potentials), while the corresponding oxidized and reduced forms (PC$^\oplus$ and PC$^\ominus$) similarly display interesting redox properties (Scheme 1.31).

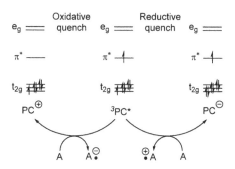

Scheme 1.31. *Redox properties of photoexcited metal polypyridyl complexes*

Taking the Ru(bpy)$_3$Cl$_2$ complex as example, the corresponding photoexcited state (^{3}[Ru(bpy)$_3$]*$^{2+}$) is both a potent reductant ($E_{\frac{1}{2}}$ ([Ru(bpy)$_3$]$^{3+}$/^{3}[Ru(bpy)$_3$]*$^{2+}$) = –0.87 V) and a good oxidant ($E_{\frac{1}{2}}$ (^{3}[Ru(bpy)$_3$]*$^{2+}$ / [Ru(bpy)$_3$]$^+$) = +0.78 V). However, the reduced form of the catalyst [Ru(bpy)$_3$]$^+$ is a strong reductant ($E_{\frac{1}{2}}$ ([Ru(bpy)$_3$]$^{2+}$/[Ru(bpy)$_3$]$^+$) = – 1.35 V) and its oxidized counterpart [Ru(bpy)$_3$]$^{3+}$ is easily reduced to its original state ($E_{\frac{1}{2}}$ ([Ru(bpy)$_3$]$^{3+}$/[Ru(bpy)$_3$]$^{2+}$) = +1.26 V) (Scheme 1.32).

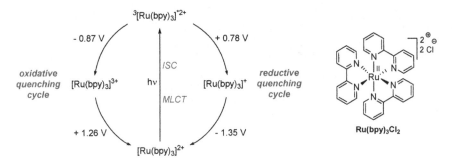

Scheme 1.32. *Redox properties of Ru(bpy)$_3$Cl$_2$ under visible light irradiation*

While in many cases the regeneration of the photoredox catalyst from its oxidized/reduced form (PC⊕/PC⊖) generally entails the use of an additional reductant or oxidant, one important feature of photoredox catalysis is that it may be possible to design processes in which a product intermediate permits it or, alternatively, it may be envisioned to take benefit from radical chain propagation mechanisms (*vide infra*). Hence, photoredox catalysis can be performed under a redox-neutral (or pseudo-redox-neutral) basis, a particular feature that renders this catalytic strategy remarkably sustainable. Moreover, while organic dyes [ROM 16] and semiconducting materials may sometimes be employed as photoredox catalysts, the comparative widespread use of metal polypyridyl complexes comes from the easy on-demand tuning of their redox properties by the flexible modulation of the polypirydyl ligands.

All in all, since its renaissance, photoredox catalysis has been established as one of the key technologies of the 21st Century and has found wonderful applications spanning from synthetic organic chemistry to polymer science

[COR 16, ZIV 16]. That being said, we may consider that the more extraordinary achievements of this catalytic method came from its synergistic association with other catalytic modes of activation including metal catalysis and organocatalysis [SKU 16, TOT 16].

1.3.2. Synthetic applications of photoredox organocatalysis

For a long time, a major challenge for photocatalysis was to be able to employ this catalytic method in an enantioselective manner. It is in this context that many strategies have been studied including the recent use of chiral metal polypyridyl complexes [WAN 15, MEG 15, BRI 15].

In 2005, Bach and co-workers reached a significant breakthrough in this field by successfully merging hydrogen bonding organocatalysis with UV-light photoredox catalysis (see section 1.3.2.3.1) [BAU 05]. Albeit it did not constitute the first example in which photochemistry had been merged with the catalytic use of a chiral organic molecule [PIV 86, PIV 90], this work certainly helped the resurgence of this particular dual catalytic strategy. Later, in 2008, this synergistic association was shown to be amenable to the use of visible light and this research field has blossomed since then [NIC 08].

As we will see in the following section, by combining the best of both worlds, a large number of original asymmetric organocatalytic reactions could be developed [BRI 16]. Indeed, while photoredox catalysis allowed innovative radical-based reactivity patterns to be unlocked, organocatalysis proved itself instrumental for the upfront induction of enantioselectivity.

For the sake of clarity, the discussion will be arranged according to which organocatalytic mode of activation was employed. As already stated, while in some cases UV-light-mediated processes will also be disclosed, our discussion will mainly focus on visible light mediated asymmetric transformations in which bond formation occurs through radical processes. Nevertheless, it is important to know that other developments include the cooperative association of these two catalytic strategies in a one-pot manner [BER 14, FEN 14, WEI 16].

1.3.2.1. *Enamine organocatalysis*

From a historical point of view, enamine organocatalysis has been one of the first organocatalytic modes of activation to be synergistically coupled with photocatalysis. In 2004, seminal studies of Córdova *et al.* showed that it is possible to realize the enantioselective α-hydroxylation of carbonyl compounds with oxygen gas under UV light irradiation in the presence of a tetraphenylporphyrin (TPP) catalyst and an amino acid organocatalyst [CÓR 04, SUN 04, IBR 06]. However, in these cases the C–O bond formation is not the result of a radical trapping of the transitory enamine and arises instead from a polar reaction mechanism in which the chiral enamine undergoes a nucleophilic attack onto singlet oxygen. Indeed, in this strategy, the PC, upon UV light photoexcitation, transfers its energy to triplet oxygen so as to *in situ* generate singlet electrophilic singlet oxygen [WAL 15]. Hence, this precedent did not take full benefit from the ability of photocatalysis to allow photoinduced electron transfers (PETs), a particular feature which MacMillan and co-workers took advantage of for the first time in 2008 in the context of the enantioselective α-alkylation of aldehydes [NIC 08], and which has known wonderful developments since then.

1.3.2.1.1. α-Alkylations of aldehydes and ketones with α-bromo carbonyl partners

As previously stated, for a long time the organocatalytic enantioselective α-alkylation of carbonyl compounds has remained troublesome. Although S_N1-based organocatalytic pathways [BRO 10], synergistic metal/organo catalysis or [AFE 16], as discussed earlier, oxidative SOMO organocatalysis could open some new perspectives in this area, the low electrophilic reactivity of alkyl halide compounds has undeniably hampered their use toward catalytically generated enamines [VIG 04, END 08, LIS 14]. It is in this particular context that a new paradigm for these kinds of transformations was born.

Taking into account that electron-rich enamines had been reported to rapidly trap electron-deficient radicals [REN 90, RUS 91], MacMillan and co-workers questioned whether it would be possible to synergistically merge asymmetric organocatalysis with the photoredox-catalyzed generation of electrophilic radicals. In 2008, they reported the proof of this concept in the context of the α-alkylation of aldehydes with electron-deficient bromo derivatives [NIC 08]. Upon irradiation with a 15 W fluorescent light bulb, it was shown that the desired α-alkylated compounds could be straightforwardly obtained in good to excellent yields and enantioselectivities when 20 mol% of organocatalyst **O6** and 0.5 mol% of Ru(bpy)₃Cl₂ were used in DMF at room temperature (Scheme 1.33). While this strategy was initially applied to the use of α-bromo diethylmalonate, it was demonstrated that, with similar efficiency, *n*-octanal could be successfully coupled with other electron-deficient bromo derivatives such as α-bromo acetophenones, 2,2,2-trifluoroethyl 2-bromoacetate, an α-bromo β-keto ester or the sterically hindered tertiary α-bromo α-methyl diethylmalonate (Scheme 1.33).

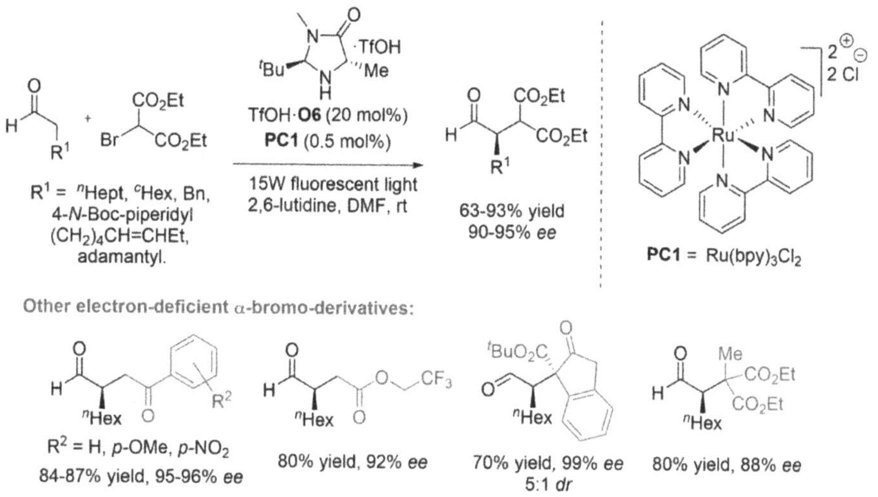

Other electron-deficient α-bromo-derivatives:

Scheme 1.33. *Proof of the synergistic asymmetric photoredox organocatalysis concept*

In order to get a better understanding of the reaction mechanism at stake in this original transformation, a series of complementary experiments and luminescence quenching studies was performed.

While the absence of fluorescent light totally inhibited the α-alkylation process, the removal of the PC from the standard protocol induced a significant decrease in the reaction efficiency since only <10% of the desired product was obtained after a 24-h period. Interestingly, in the absence of Ru(bpy)₃Cl₂, the productivity of this organocatalytic process could be maintained when a UV irradiation source (300–350 nm) was used. This result suggests that an alternative photoinduced reaction mechanism could be at stake, for which the authors invoked that the generation of the alkyl radical could come from a photolytic bond homolysis.

With respect to the action of the PC, although it had been previously shown that the photoexcited ^{3}[Ru(bpy)$_3$]*$^{2+}$ (PC*) could efficiently reduce the electrodeficient bromonitromethane following an oxidative quench [OIS 78], a decrease in PC* luminescence was not observed in the presence of increasing concentrations of diethyl bromomalonate or α-bromo acetophenone. Hence, it negated the possibility that PC* acted as a reductant in this photoredox organocatalytic process and led the authors to propose that this reduction event was instead promoted by the reduced PC (PC$^\ominus$ = [Ru(bpy)$_3$]$^+$). This suggestion was consistent with their observation that a O6-derived enamine (generated in stoichiometric quantities) decreased the emission intensity of PC* (Stern–Volmer quenching constant of 10 M^{-1}), and thus that PC* most probably serves as an oxidant to initiate this reaction. Last but not least, it was demonstrated that the enamine does not follow a SOMO pathway (intermediary formation of a 3π-electron radical cation) and directly participates as a SOMOphile in the key C–C bond forming step. Indeed, a radical clock experiment employing racemic 2-phenylcyclopropyl acetaldehyde led cleanly to the corresponding diastereoisomeric alkylation products without undergoing any cyclopropyl ring opening side processes (Scheme 1.34).

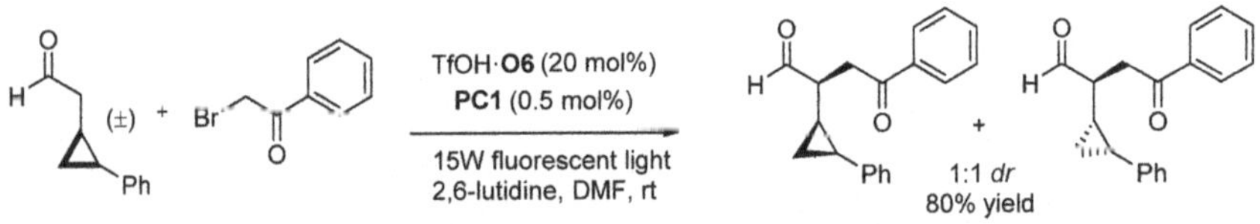

Scheme 1.34. *Radical clock control experiment*

With all these observations, MacMillan and co-workers formulated a mechanistic proposal in which two interwoven catalytic cycles operate in an overall redox-neutral manner (Scheme 1.35).

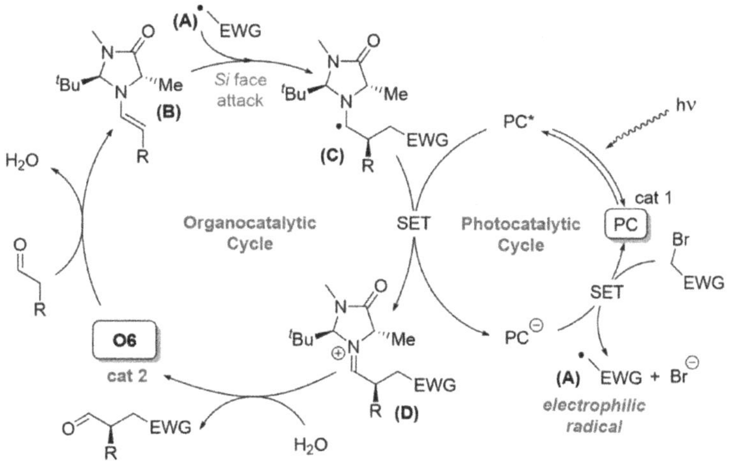

Scheme 1.35. *MacMillan's mechanistic proposal for this transformation*

According to this, the reduced photoredox catalyst (PC⊖) (which is a strong reductant and could initially be formed by PET between PC* and a sacrificial quantity of enamine (B)) would reduce the halide derivative into the corresponding electrophilic alkyl radical (A), which would then preferentially couple with the less sterically hindered *Si* face of the *in situ* formed SOMOphilic enamine (B) [ZHA 16]. The resulting α-amino radical (C) was proposed to be easily oxidized by PC* into iminium (D), a process which would close both catalytic cycles by allowing both the regeneration of PC⊖ and, after hydrolysis, that of the amine organocatalyst.

Since then, this ideal mechanistic picture has evolved somewhat, notably with the work of Yoon and co-workers [CIS 15]. Indeed, while it appears kinetically unlikely that a weak concentration of the α-amino radical (C) ensures the rapid reductive quenching of the transient excited photocatalyst PC*, Yoon and co-workers alternatively proposed the involvement of a radical chain propagation mechanism in which the α-amino radical (C) directly reduces the electron-deficient bromo derivative (Scheme 1.36); an active reductive participation of α-amino radical often underestimated in photoredox catalysis [ISM 13]. This hypothesis could be confirmed by the measurement of the quantum yield of this α-alkylation process (Φ = (number of molecule produced)/(number of photons absorbed)) that, with a value of 18, is far superior to the maximal value Φ = 1 that would be expected if no propagation occurred.

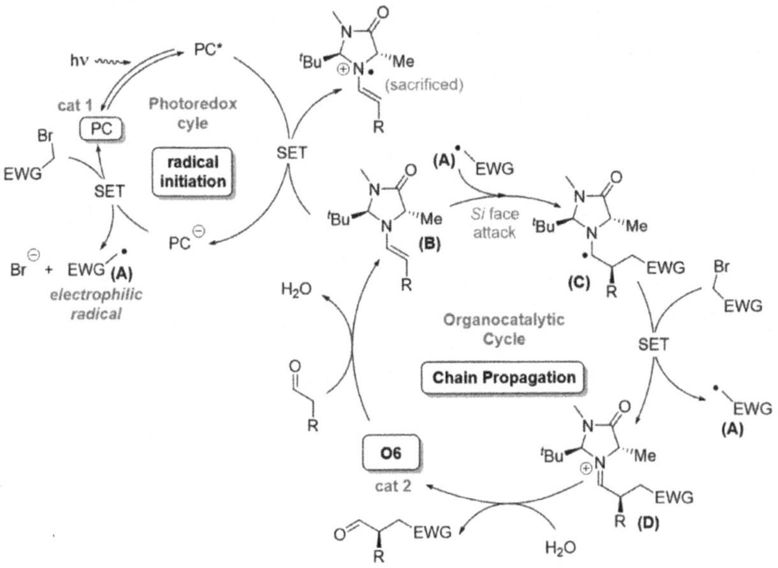

Scheme 1.36. *Yoon's radical chain propagation mechanism*

Considering that, under the optimized reaction conditions, Yoon and co-workers could only observe a 10% diminution of the phosphorescence of the PC, it indicated that only a very limited portion of the PC initiates the α-alkylation process and, thus, that radical chain lengths are in fact quite long (with a lower limit of 180) [CIS 15]. Therefore, the actual role of the PC is to continuously reinitiate the radical chain propagation upon its termination rather than ensuring the formation of the electrophilic radical partner for every single organocatalytic cycle.

On the other hand, Yoon and co-workers directly put the limited efficiency of the radical initiation step in relation with the very small available concentration of enamine (B) and the restricted capacity of the latter to induce the reductive quench of PC* (the Stern–Volmer quenching constant of enamine (B) is estimated at 10 M^{-1}). In this context, Yoon and co-workers proposed that the additional use of a more potent sacrificial reductant should enhance the reaction rate by easily promoting the formation of PC $^{\ominus}$. This strategy, which is still poorly employed at the moment, revealed itself particularly fruitful since only adding 0.5 mol% of *N,N*-dimethyl toluidine, a particularly good reductive quencher of $^3[Ru(bpy)3]^{*2+}$, [BOC 79] sped up the rate of α-alkylation of *n*-octanal with the α-bromo diethylmalonate by approximately one order of magnitude (Scheme 1.37) [CIS 15].

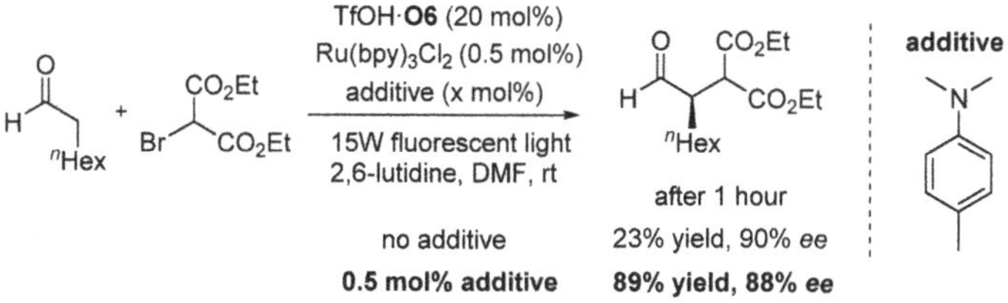

Scheme 1.37. *Yoon's modification of the α-alkylation of aldehydes*

Alternative photoredox catalysts

Due to the limited availability and potential toxicity of precious metals such as ruthenium and iridium upon which a vast majority of photoredox catalytic transformations are based, several research groups questioned whether cheaper and more environmentally friendly alternatives could allow this photoredox organocatalytic α-alkylation process.

Studying the use of earth-abundant first row transition metal complexes, Ceroni and co-workers reported that the homogeneous iron(II) complex Fe(bpy)$_3$Br$_2$ (**PC2**) allowed the desired reactivity to take place [GUA 15]. Under reaction conditions almost identical to those developed by MacMillan and co-workers, it was necessary to raise the amount of PC to 2.5 mol% so as to attain comparable reaction rates, yields and enantioselectivities (Scheme 1.38).

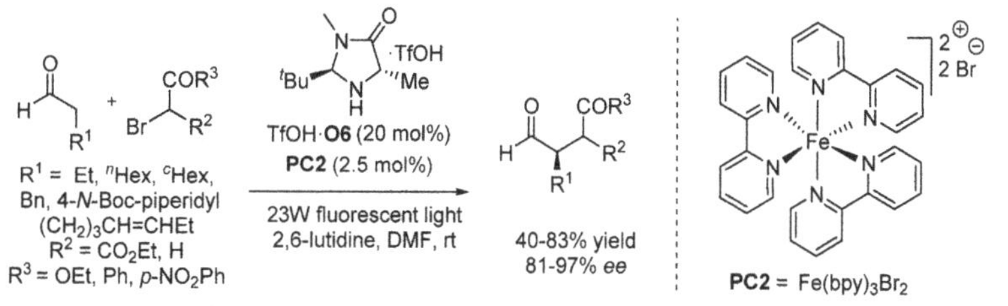

Scheme 1.38. *Fe(bpy)$_3$Br$_2$-catalyzed synergistic α-alkylation of aldehydes*

One particularly interesting feature of this iron-mediated transformation is that Fe(bpy)$_3$Br$_2$ is typically not a good candidate for dynamic electron-transfer processes because of the extremely short lifetime of its lowest energy excited state [CAN 10]. Yet, detailed mechanistic studies including spin-trapping and radical clock experiments confirmed the generation of alkyl radicals from the starting bromo partner and that the transitory enamine does behave as a SOMOphile. While it appeared clear that Fe(bpy)$_3$Br$_2$ did played the role of a photoredox catalyst in this transformation, the relative efficiency of the α-alkylation process was, in correlation to Yoon's demonstration, attributed to the involvement of a radical chain propagation mechanism. Hence, this work demonstrated that, in such photoredox organocatalytic process, the redox efficacy of the PC is not primary as long as it is sufficient to induce radical initiation.

Following the same goal to use more eco-friendly photocatalysts, Zeitler and co-workers, followed by Ferroud et al, considered the development of entirely metal-free strategies [NEU 11, FID 12]. While the ability of organic dyes to promote visible-light PETs has been known for several decades, the resurgence of metal-based photoredox catalysis has surely encouraged the field of organic photoredox catalysis to blossom [ROM 16]. In this particular context, Zeitler's group envisioned the use of organic dyes in the α-alkylation of aldehydes [NEU 11]. The examination of various red and orange dyes with a green LED light source (λ = 530 nm) led to the identification of eosin Y (**PC3**) as a competent PC for this transformation (Scheme 1.39). While good yields and enantioselectivities could be obtained at −5 °C, the coupling process could also be performed under direct sunlight irradiation at 30 °C at the cost of small erosion of the enantiocontrol. From a mechanistic perspective, the authors proposed that eosin Y and Ru(bpy)$_3$Cl$_2$ behave in a similar fashion. Even though the authors did not mention it, it seems reasonable that a radical chain propagation process is involved.

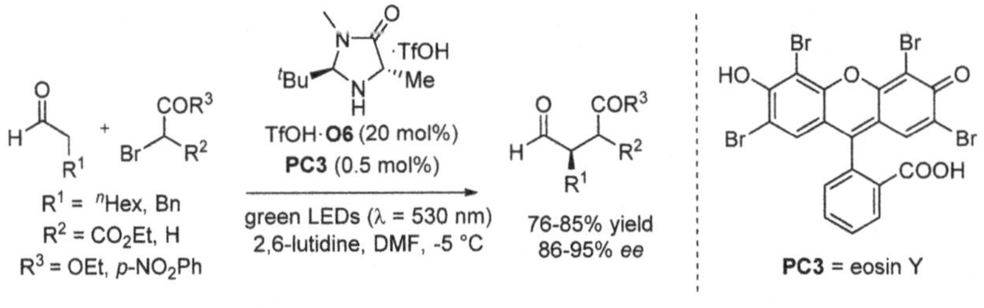

Scheme 1.39. *Purely organic photoredox organocatalytic α-alkylation of aldehydes*

In line with this work, Zeitler and co-workers showed that this entirely organic photoredox organocatalytic system could be successfully applied to microfluidic synthesis [NEU 12], while Ferroud and co-workers showed that employing Rose Bengal, in DMSO in the presence of catalytic quantity of LiCl, allowed the desired aldehyde α-alkylation to take place at room temperature with limited irradiation times and reduced organocatalyst loading [FID 12].

As a continuation of this search for alternative photocatalysts, it was also independently demonstrated by the groups of König and Pericàs that heterogeneous semiconductors such as nanocristalline PbBiO₂Br and non-toxic bulk Bi₂O₃, which possess moderate to low band gaps (energy difference between the top of the valence band and the bottom of the conduction band), could similarly induce the desired reactivity [CHE 12, RIE 14]. Such a kind of heterogenous photoredox strategy was also developed in the context of chiral metal-organic frameworks by He and co-workers [WU 12].

Alternative class of substrates

In 2014, as a continuation of their interest in the organocatalytic activation of β-carbonyl compounds [XU 14], Luo and co-workers reported the α-alkylation of this particular class of compounds by merging photoredox catalysis with primary amine organocatalysis. In more detail, β-keto esters, 1,3-diketones and β-ketoamides were shown to enantioselectively couple with various phenacyl bromide derivatives by merging the catalytic use of Ru(bpy)₃Cl₂ (**PC1**, 1 mol%) with that of the diamine **O8** (20 mol%). Using a 33 W compact fluorescent light source, and NaHCO₃ as base in acetonitrile at room temperature, a wide variety of alkylated products bearing an all-carbon quaternary stereogenic center was obtained in good to excellent yields, while the specific use of *N*-aryl ketoamides led to the diastereoselective formation of the corresponding hemiaminal products (Scheme 1.40) [ZHU 14].

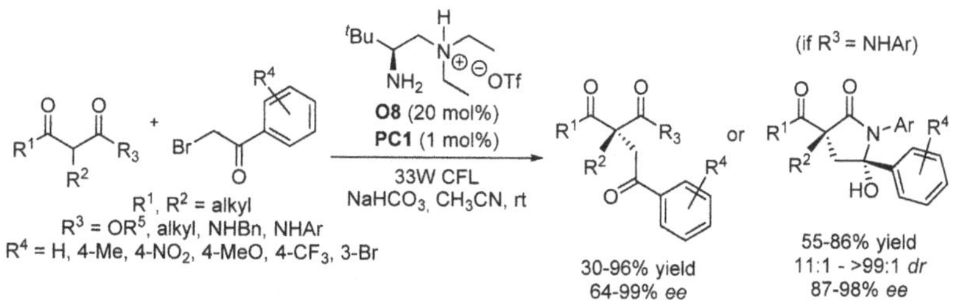

Scheme 1.40. *Photoredox organocatalytic α-alkylation of β-carbonyl compounds*

Several control experiments showed the critical role of NaHCO$_3$ for reaching high productivity and the requirement of light irradiation for this reaction to proceed. Interestingly, in parallel to MacMillan's work [NIC 08], the absence of PC Ru(bpy)$_3$Cl$_2$ still allowed the reaction to occur, albeit in lower yield. It indicated that an alternative photoinduced mechanism could competitively be at stake, although it remained minor compared to that of the **PC1**-catalyzed process. The authors attributed this observation to the partial involvement of an electron donor acceptor mechanism as reported by Melchiorre and co-workers (*vide infra*) [ARC 13, ARC 14, BAH 16], although no direct evidence was given to support this hypothesis.

It was proposed that the PC promotes the reduction of phenacyl bromide to the corresponding alkyl radical and that the latter stereoselectively couples with the *Re* face of the **O8**-derived enamine by means of a key H-bonding network (Scheme 1.41) [ZHA 16]. The question of the intervention of a radical chain propagation process was not raised.

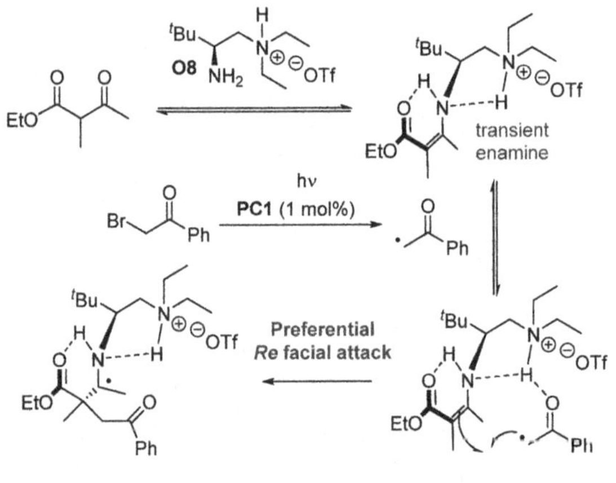

Scheme 1.41. *Luo's model for the **O8**-controlled enantioinduction*

Alternative photoinduced processes exempt of the need for an external photoredox catalyst

All the α-alkylation reactions discussed so far were based on the synergistic use of a photoredox catalyst with an amine-based organocatalyst. However, as discussed earlier, the removal of the PC from the catalytic systems did not always wholly suppress the desired alkylation process [NIC 08, ZHU 14]. This intriguing observation hinted that other photoinduced mechanisms could very well be at stake, such that the field of photo-organocatalysis is undeniably more intricate than it may seem.

Since 2013, Melchiorre and his group have been involved in the discovery of such photoredox catalyst-free photoinduced organocatalytic transformations. To this end, Melchiorre and co-workers capitalized on the intrinsic photochemical activities of chiral enamines, either by means of photoactivation of electron donor–acceptor (EDA) complexes or by their direct photoexcitation [ARC 13, ARC 14, SIL 15, BAH 16].

EDA complexes

As formulated more than 50 years ago by Mulliken [MUL 50], an electron-rich compound possessing a low IP (donor) may interact with an electron-poor molecule (acceptor) and form, even in solution, the corresponding intimate EDA complex (also referred to as charge–transfer [CT] complex). This ground state association [D,A] is typically characterized by the formation of a new weak absorption band, the CT band, which is not observed in the UV–visible spectra of the individual components and is generally situated at higher wavelengths (redshifted). When the CT band lies in the visible-light domain, the formation of an EDA complex can straightforwardly be observed by a characteristic strong coloration of the reaction mixture.

An important characteristic of EDA complexes is that, upon irradiation at the wavelength of its CT band, the EDA complex may reach an excited state [D,A]* susceptible to trigger, within the solvent cage, an internal electron transfer generating the radical ion pair [D$^{\bullet\ominus}$,A$^{\bullet\oplus}$]. The latter can then evolve through different pathways (rearrangements, additions, eliminations, etc.), undergo a back-electron transfer or, alternatively, escape the solvent cage to induce radical processes [MUL 52, FOS 80, MOR 13]. Hence, a light-promoted electron transfer does not always entail the use of a PC, especially when the generation of an EDA complex is possible. Interestingly, although the physicochemical properties of EDA complexes have been extensively studied, their synthetic applications have, until quite recently, been sporadically reported [LIM 16].

In 2013, based on few literature precedents [CAN 77, RUS 91], Melchiorre and co-workers recognized that, due to their electron richness, enamines are potentially good donors. Therefore, upon the selection of a suitable electron-poor acceptor, it was anticipated that an *in situ* generated chiral enamine could promote, via the formation of an EDA complex, the required photochemical activation and, thereby, could offer the possibility to develop asymmetric photo-organocatalytic transformations without the need for an external photosensitizer.

The proof of this concept was initially achieved when employing the very much electron-deficient 2,4-dinitrobenzyl bromide as acceptor, n-butanal (3 equiv.), organocatalysts O9-O11 (20 mol%) and a 23 W compact fluorescent light source. Under these conditions, Melchiorre and co-workers demonstrated that the corresponding enantioselective α-benzylation reaction straightforwardly occurred in good yields in the absence of any photoredox catalyst, the best enantioselectivity being obtained with O11 (Scheme 1.42) [ARC 13].

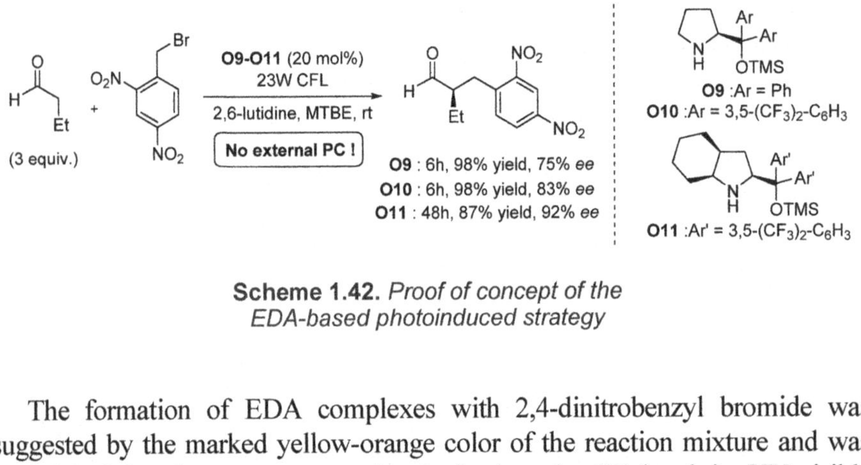

Scheme 1.42. *Proof of concept of the EDA-based photoinduced strategy*

The formation of EDA complexes with 2,4-dinitrobenzyl bromide was suggested by the marked yellow-orange color of the reaction mixture and was established by the appearance of a bathochromic CT band in UV-visible spectroscopy. While, as we will discuss later in more detail (section 1.3.2.1.2), this strategy allowed the enantioselective α-benzylation of a larger range of carbonyl compounds to be realized [ARC 13, ARC 14], it was shown that phenacyl bromides were also particularly adequate electron acceptors. Indeed, such electron-poor partners, which similarly led to the formation of EDA complexes with O11-derived enamines, were shown to equally promote the corresponding α-alkylation of aldehydes in good yields and good enantioselectivities under photoredox catalyst-free reaction conditions (Scheme 1.43).

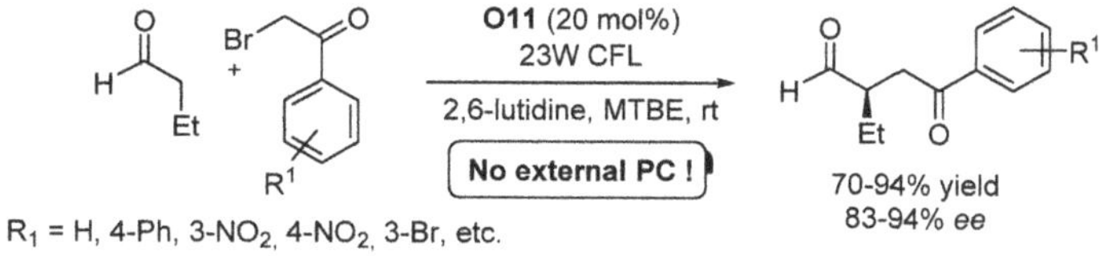

Scheme 1.43. *Photo-organocatalytic α-alkylation of aldehydes based on the EDA strategy*

The akin α-alkylation of enals was operated with complete γ-selectivity, although in these cases lower enantioselectivities were attained [ARC 13]. On the other hand, cyclohexanone was also shown to enantioselectively couple with phenacyl bromides, provided that the primary amine organocatalyst **O12** (20 mol%) and TFA (40 mol%) were used under the irradiation of a 300 W xenon lamp (300–600 nm) (Scheme 1.44) [ARC 14].

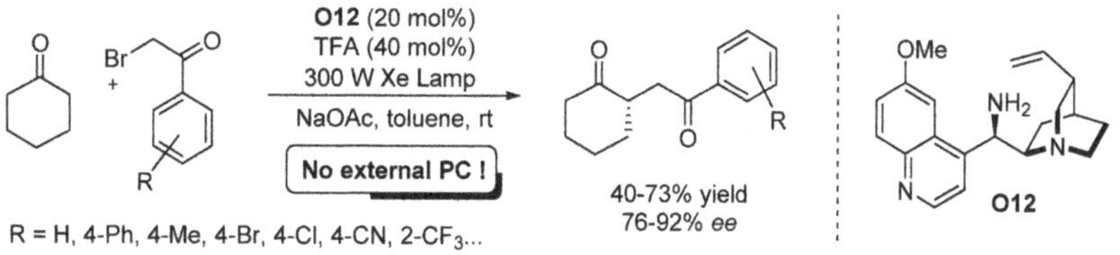

Scheme 1.44. *Photo-organocatalytic α-alkylation of cyclohexanone*

In this case, the need for a more powerful light source may indicate that the formation of an EDA complex is less marked or that the CT band is less redshifted as a result of the reduced electron richness of the secondary enamine. Alternatively, with such a powerful light source, the direct photoexcitation of the enamine cannot be fully ruled out (*vide infra*).

From a mechanistic perspective, the authors initially proposed that, upon the formation of an enamine by condensation of the aldehyde (ketone) partner with the organocatalyst, the colored enamine/alkyl bromide (phenacyl bromide derivative or 2,4-dinitrobenzyl bromide) EDA complex was formed. Its photoexcitation by visible light would allow it to reach an excited state, potentially evolving by an SET from the donor to the acceptor. The resulting radical ion pair was then envisioned to evolve via the departure of the bromide anion followed by an in-cage stereoselective radical combination (Scheme 1.45).

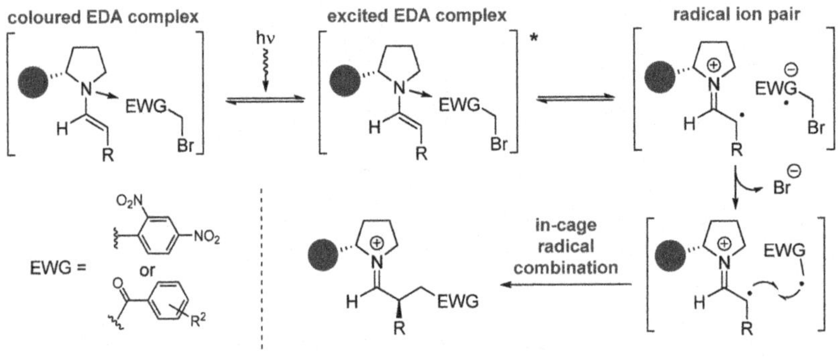

Scheme 1.45. *Initial mechanistic proposition for the EDA photo-organocatalytic strategy*

However, further studies revealed a different mechanistic scenario for such EDA-based photo-organocatalytic alkylation processes [BAH 16]. Indeed, whether 2,4-dinitrobenzyl bromide or phenacyl bromide was used, the **O10**-catalyzed α-functionalizations of *n*-butanal were estimated to operate with high quantum yields ($\Phi \approx 30$), incompatible with an in-cage radical combination ($\Phi_{max} = 1$) and unmistakably showing that radical chain propagation occurs. Hence, the photoexcitation of the EDA complex was alternatively proposed to serve the radical initiation process, much like what Yoon's group observed when a photoredox catalyst (Ru(bpy)$_3$Cl$_2$) is used [CIS 15]. Rather than undergoing in-cage radical combination, Melchiorre and co-workers proposed that the liberation of the radical **A** out of the solvent cage allowed to initiate (or reinitiate) the propagation process at the cost of sacrificing some of the starting aldehyde and organocatalyst (Scheme 1.46).

In the same mechanistic study [BAH 16], a deeper investigation allowed to raise that the nature of the key propagation step (**C** + EWG-CH$_2$Br ⟶ **D** + EWG-CH$_2$•) may depend on the electrophilic partner. Realizing cyclic voltammetry studies on a model iminium, the redox potential of the **D/C** couple was estimated at –0.95 V versus Ag/Ag$^+$, which comforted that the reduction of 2,4-dinitrobenzyl bromide ($E_p^{red} = -0.66$ V) is thermodynamically favored and, as proposed by Yoon and others, can occur via an SET manifold. However, in the case of phenacetyl bromide ($E_p^{red} = -1.35$ V), the direct reduction process should not be spontaneous and was proposed to preferentially occur via a bromine transfer followed by a polar elimination of a bromide anion (Scheme 1.47).

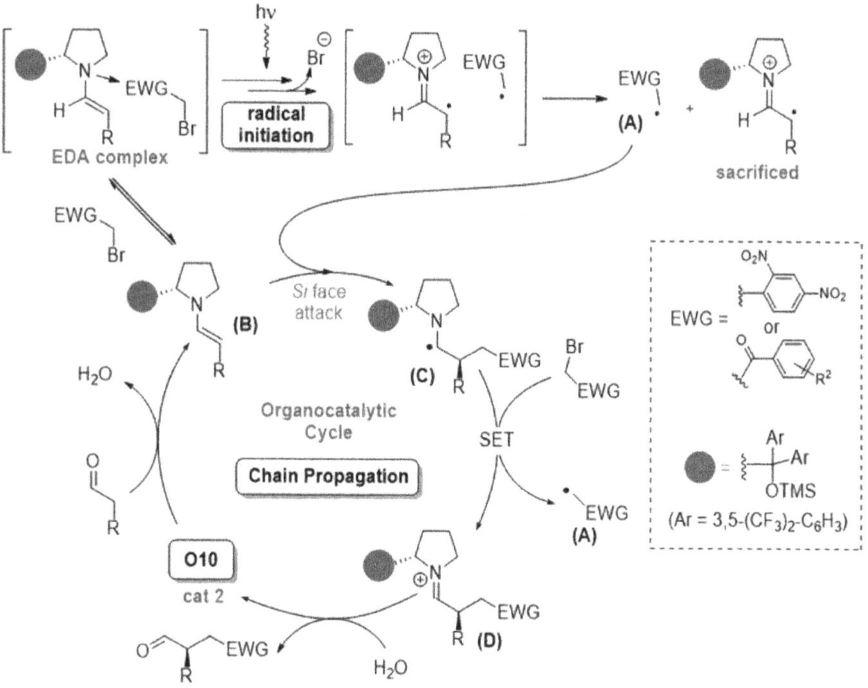

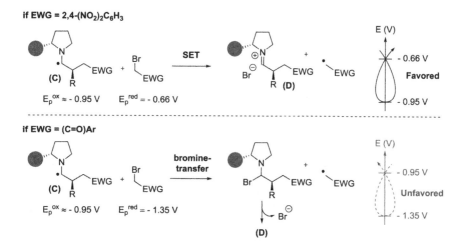

Scheme 1.46. *Revised mechanistic proposition for the EDA photo-organocatalytic strategy*

if EWG = 2,4-(NO₂)₂C₆H₃

$E_p^{ox} \approx -0.95$ V $E_p^{red} = -0.66$ V

-0.66 V **Favored**
-0.95 V

if EWG = (C=O)Ar

$E_p^{ox} \approx -0.95$ V $E_p^{red} = -1.35$ V

-0.95 V
-1.35 V **Unfavored**

Scheme 1.47. *Mechanistic insights regarding the key propagation step*

Direct photoexcitation of enamines

As a continuation of their work on the photo-organocatalytic α-alkylation of aldehydes, Melchiorre and co-workers investigated whether the EDA activation strategy could be applied to the use of tertiary and quaternary α-bromo malonates [SIL 15].

Satisfyingly, upon using *ent*-**O10** (20 mol%) in MTBE at room temperature under the irradiation of a 23 W compact fluorescent lamp (CFL), various aldehydes smoothly underwent the desired α-functionalization reaction. While the corresponding α-alkyl aldehydes were typically obtained with good yields and excellent enantioselectivities, the use of α-aryl enals (with **O10**) was allowed to regioselectively obtain the corresponding γ-functionalized products with good enantiocontrol (Scheme 1.48).

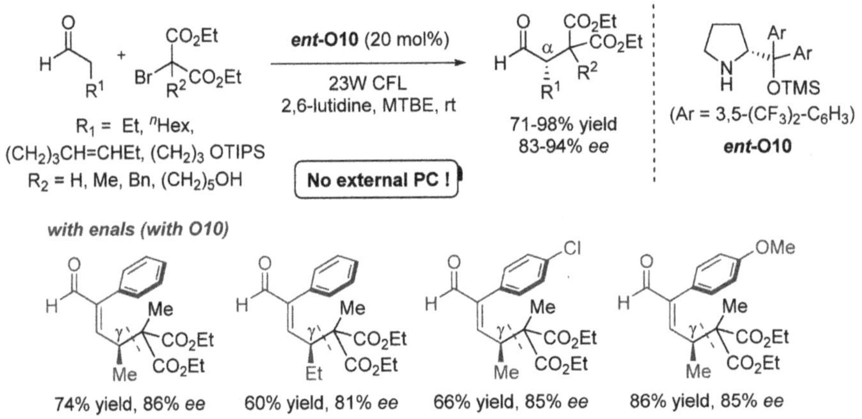

Scheme 1.48. *Light-induced α-alkylation of aldehydes and enals with α-bromo malonates*

Even though no extra photoredox catalyst had to be used, the careful exclusion of light completely suppressed the alkylation process, proving the photochemical nature of the reaction. In order to gain more insights into this peculiar photoredox catalyst-free photoinduced organocatalytic transformation, several control experiments were performed.

The presence of 1 equiv. of TEMPO completely inhibited the reaction, which confirmed the radical nature of the mechanism. On the other hand, the formation of a ground state EDA complex was excluded as no CT absorption band was observed for the reaction mixture. Hence, a completely unforeseen photochemical process was operating.

When the reaction was performed with a cut-off filter at 385 nm (with which only wavelengths with a $\lambda \geq 385$ nm go through), the reaction occurred with the same efficiency, meaning that whatever photoactive species induced this transformation had to absorb in this wavelength region. A careful look at the absorption spectrums showed that the only photoabsorbing compound in this domain is the transitory **O10**-derived enamine, suggesting that it does play the role of photoinitiator. This result was further supported by Stern–Volmer studies showing that an excited **O8**-derived enamine, prepared from 2-phenyl acetaldehyde, was effectively quenched by the α-bromo diethyl malonate partner.

In line with Yoon's work, Melchiorre and co-workers also proposed a radical chain mechanism for this reaction, in which sacrificial quantities of the enamine **B**, upon photoexcitation, serve to initiate (or reinitiate) the formation of the necessary alkyl radicals **A**. This hypothesis was later confirmed by the determination of a high quantum yield ($\Phi_{estimated} = 182$) [BAH 16]. It was proposed that the bulky diarylsilyloxy group shields the *Si* face of the enamine **B** such that the coupling between the alkyl radical **A** and the enamine **B** preferentially occurs from its less hindered *Re* face (Scheme 1.49). As for the key propagation step, considering the important reduction potential of α-bromo diethylmalonate ($E_p^{red} = -1.69$ V), Bahamonde and Melchiorre suggested that it should occur according to a bromine-transfer process rather than a direct SET [BAH 16].

In addition to the synthetic utility of harnessing the photochemical activity of enamines for the development of new photo-organocatalytic processes, this work shed some original interesting perspectives in the realm of photo-organocatalytic transformations. From an original vision in which a photoredox catalyst was needed to perform the desired α-alkylation reactions, this field progressively recognized the prominent contribution of radical chain propagation mechanisms while other photoactivation strategies may also take part. Considering the polychromatic variety of the light source that have been used as well as the variance in electron richness of the transitory enamines [MAY 12, LAK 12], the differentiation of the reaction mechanisms that are at stake may not always be clear-cut. Hence, it is important to acknowledge the complexity of photo-organocatalysis and that this field, which is still in its infancy, will continue to bring more and more mechanistic insights along its foreseeable growth. For these reasons, it is important for the reader to consider that, in this chapter, we present the mechanistic propositions of the authors but we cannot entirely warrant their accuracy.

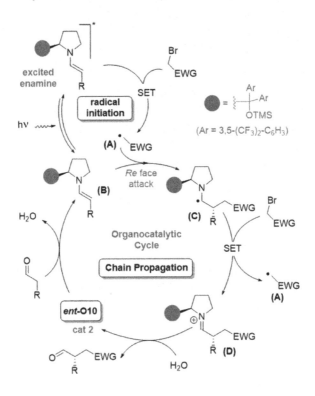

Scheme 1.49. *Mechanistic proposal for the stereoselective α-alkylation of aldehydes based on the photochemical activity of enamines*

1.3.2.1.2. α-Benzylation of aldehydes

In 2010, as a continuation of their work on the photoredox organocatalytic asymmetric α-alkylation of aldehydes with α-bromo keto derivatives [NIC 08], MacMillan and co-workers reported the corresponding α-benzylation reaction [SHI 10]. Submitting a set of aldehydes to a variety of electron aryl and heteroaryl methylene bromides in the presence of TfOH•O13 (20 mol%) and the *fac*-Ir(ppy)₃ photocatalyst (**PC4**, 0.5 mol%) in DMSO at room temperature, the corresponding α-benzylated products were reliably obtained in good yields and enantioselectivities (Scheme 1.50).

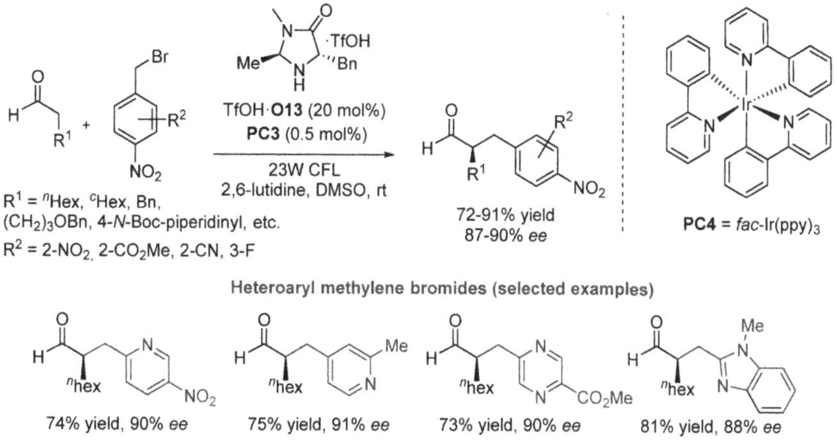

Scheme 1.50. *MacMillan's α-benzylation of aldehydes
by photoredox organocatalysis*

Upon performing Stern–Volmer studies (fluorescence quenching studies) with **PC4**, it could be demonstrated that, in opposition to their previous study based on the use of $Ru(bpy)_3Cl_2$,[NIC 08], **PC4*** was efficiently quenched by 2,4-dinitrobenzyl bromide rather than by the enamine, suggesting that the reduction step directly occurs from the PC-excited state to the benzylic bromo derivative. This observation is totally in accord with the important reductive aptitude of **PC4*** ($E_{\frac{1}{2}} = -1.73$ V vs. SCE) and the electron-deficient character of 2,4-dinitrobenzyl bromide. On the other hand, it may explain the inaptitude of this catalytic system to promote the desired benzylation reaction with benzyl bromide. Indeed, such electron-neutral benzylic bromides are much more difficult to reduce ($E_{\frac{1}{2}}$ (BnBr) $= -1.85$ V vs. SCE), which probably hampers the requisite reduction process. On the other hand, it is significant that the reduction of basic heteroaryl bromides was facilitated by their protonation with additional hydrogen bromide (HBr), as a way of further diminishing their electron richness and ease of the electron transfer.

With these mechanistic studies, MacMillan and co-workers proposed the following intertwined catalytic cycles in which the excited photocatalyst **PC4*** would promote the formation of the benzylic radical **A**, inclined to preferentially couple with the *Si* face of enamine **B**. The resulting α-amino radical **C** was suggested to allow the regeneration of the PC by SET, together with the formation of iminium **D**. The latter, upon hydrolysis, would then afford the desired α-benzylated aldehyde and permit the release the organocatalyst (Scheme 1.51).

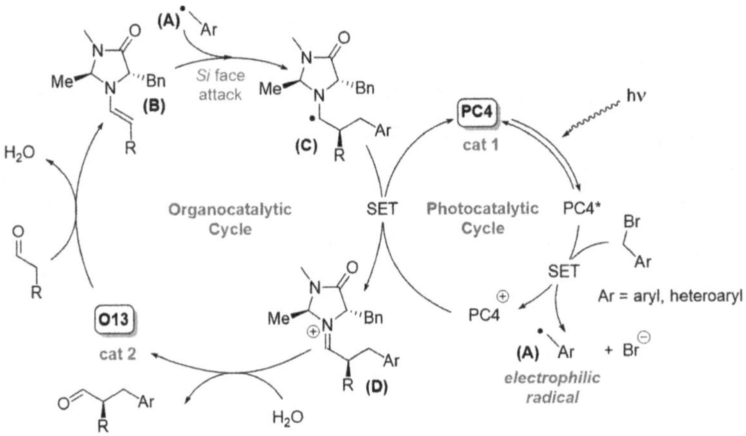

Scheme 1.51. *MacMillan's mechanistic proposition for the α-benzylation of aldehydes by photoredox organocatalysis*

While this mechanistic picture is in agreement with their previous report on the synergistic photoredox organocatalytic α-alkylation of aldehydes [NIC 08], a radical chain propagation is likely to participate in this transformation when considering Melchiorre's subsequent work in the field [ARC 13, ARC 14, BAH 16].

As already mentioned, Melchiorre and co-workers reported that the akin asymmetric α-benzylation of aldehydes could be performed by capitalizing on the aptitude of tertiary enamines to form EDA complexes with electron-poor benzylic bromides (Scheme 1.52) [ARC 13].

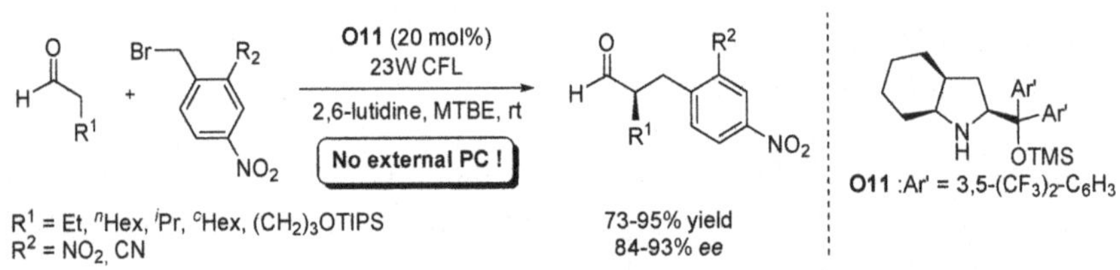

R^1 = Et, ^{n}Hex, ^{i}Pr, ^{c}Hex, (CH$_2$)$_3$OTIPS
R^2 = NO$_2$, CN

73-95% yield
84-93% ee

Scheme 1.52. *Melchiorre's α-benzylation of aldehydes based on the EDA strategy*

While this photo-organocatalytic method could later be extended to the use of cyclic ketones by using the quinidine-derived organocatalyst **O12** [ARC 14], detailed mechanistic studies revealed that a radical chain propagation mechanism is operating as a result of the aptitude of α-amino radicals to directly reduce electron-poor benzylic bromides (Scheme 1.46) [BAH 16].

1.3.2.1.3. α-Trifluoro- and perfluoroalkylation of aldehydes

Taking into account the widespread applications of fluorine compounds, MacMillan and co-workers questioned whether the photoredox organocatalysis paradigm could open the way to the enantioselective α-trifluoromethylation of aldehydes. In 2009, they demonstrated that the desired α-functionalization reaction can occur when submitting various aldehydes to trifluoromethyl iodide in the presence of **O6•TFA** (20 mol%) and Ir(ppy)₂(dtb-bpy)PF₆ **(PC5)** in DMF at –20°C. Under these reaction conditions, a variety of α-trifluoromethyl aldehydes, as well as several perfluoroalkylated counterparts, were obtained in good yields and almost perfect stereocontrol (Scheme 1.53) [NAG 09].

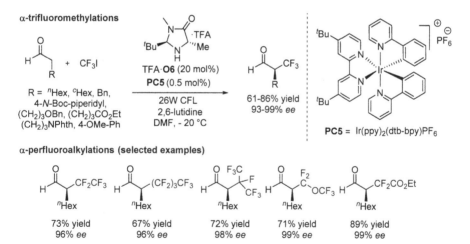

Scheme 1.53. *α-Trifluoro- and perfluoroalkylations of aldehydes by synergistic catalysis*

The yields and *ee*s reported were those obtained after the *in situ* reduction into the corresponding alcohols, as a probable way to prevent any epimerization to take place.

While this reaction could also proceed using Ru(bpy)$_3$Cl$_2$ as PC, better yields were typically obtained with **PC5**. Stern–Volmer quenching studies allowed to establish that, similarly to the Ru(bpy)$_3$Cl$_2$-catalyzed α-alkylation of aldehydes [NIC 08], the excited photocatalyst **PC5*** underwent a reductive quench with sacrificial quantities of enamines so as to, in turn, allow the generation of electrophilic fluoroalkyl radicals. In this context, MacMillan and co-workers proposed a reaction mechanism close to that formerly reported (Scheme 1.45). Note that the related photoinduced coupling of pyrrolidine-derived enamines with perfluoroalkyl iodides had been previously reported, for which the formation of EDA complexes and the involvement of a radical chain propagation mechanism were proposed [CAN 75, CAN 77].

1.3.2.1.4. α-Cyanoalkylation of aldehydes

As a direct continuation of their interest in the α-functionalization of aldehydes, MacMillan and co-workers explored other sources of electrophilic radicals. Taking into account the rather low reduction potential of α-bromoacetonitrile ($E_{1/2}^{red} \approx -0.69$ V vs. SCE in DMF), the authors envisioned that this particular bromo derivative should straightforwardly generate the corresponding cyanoalkyl radical upon reductive quenching of the Ru(bpy)$_3$Cl$_2$ excited state (**PC1***). Under photoredox organocatalysis, it was anticipated that such electrophilic radicals should open the way to the enantioselective α-cyanoalkylation of aldehydes. This strategy met the anticipated success, the coupling of a variety of aldehydes with α-bromoacetonitrile leading efficiently to the corresponding enantioenriched β-cyano aldehydes when submitted to organocatalyst **O6** (20 mol%) and Ru(bpy)$_3$Cl$_2$ (1 mol%) in DMSO at room temperature under the irradiation of a CFL (Scheme 1.54) [WEL 15].

The use of tertiary and quaternary cyanoalkyl bromides was also amenable, although poor diastereoselectivities were typically obtained (*dr* < 3:1) and better results were obtained when using organocatalyst **O14**. The authors suggested that the enhanced catalytic activity of **O14** compared to **O6** may come from the improved aptitude of **O14** to generate the corresponding enamine, as proven by NMR measurements of the relative equilibrium constants for enamine formation ($K_{O14} > K_{O6}$).

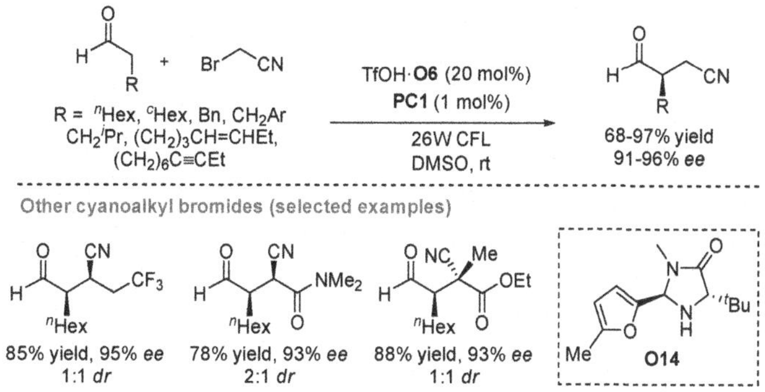

Scheme 1.54. *α-Cyanoalkylation of aldehydes by photoredox organocatalysis*

While the mechanistic proposal for this transformation remains in line with that of the α-alkylation of aldehydes with α-bromo malonates (Scheme 1.55), once again a radical chain propagation mechanism, in which intermediate **C** directly reduces the cyanoalkyl bromide into the corresponding electrophilic radical **A**, cannot be fully ruled out.

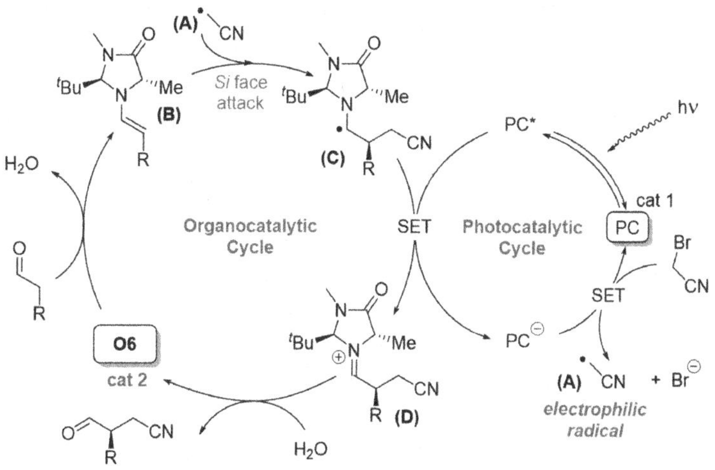

Scheme 1.55. *MacMillan's mechanistic proposal for the α-cyanoalkylation of aldehydes*

1.3.2.1.5. α-Oxyamination of aldehydes

In 2009, Koike and Akita reported the photoinduced catalytic oxyamination of aldehydes by merging enamine organocatalysis with photoredox catalysis [KOI 09]. Submitting several aldehydes to TEMPO in the presence of morpholine (20 mol%) and $Ru(bpy)_3(PF_6)_2$ (**PC6**, 2 mol%) in acetonitrile allowed, under the irradiation of a Xe lamp ($\lambda > 420$ nm), to obtain the corresponding α-oxyaminated products in average to good yields (Scheme 1.56).

Note that this reaction did not work with secondary aldehydes, such as cyclohexane carboxaldehyde, or with cyclic ketones (cyclohexanone), justifying the need for an efficient enamine formation. Indeed, the direct use of the cyclohexanone/morpholine enamine did afford the desired α-oxyamination product, when submitted to TEMPO and **PC6** (not shown).

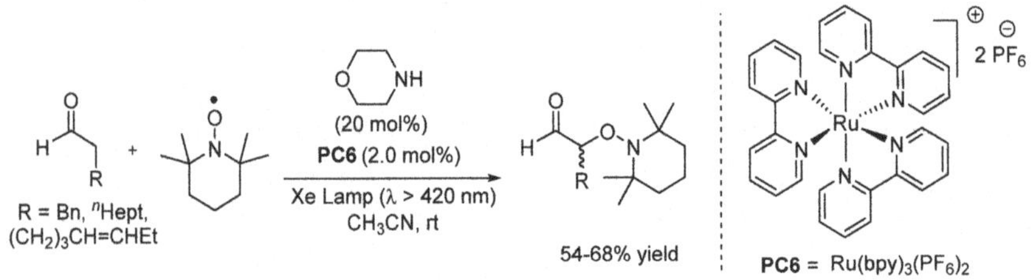

Scheme 1.56. *Photoredox organocatalytic oxyamination of aldehydes*

For this transformation, Koike and Akita proposed that the PC-excited state **PC6***, upon reductive quench, would promote the oxidation of the transitory enamine **A** into the corresponding radical cation **B**. The latter was suggested to undergo a radical–radical coupling with TEMPO to afford the iminium **C** which would, upon hydrolysis, regenerate the morpholine organocatalyst and deliver the α-oxyaminated aldehyde (Scheme 1.57).

Yet, as acknowledged by the authors, this mechanistic picture does not explain how the reduced photocatalyst **PC6** ⊖ is subsequently oxidized so as to allow the regeneration of the PC. Moreover, while **PC6*** is indeed reduced by the enamine, quenching experiment studies also showed that it is also the case with TEMPO. Hence, further mechanistic studies would have been necessary to fully understand how this transformation operates.

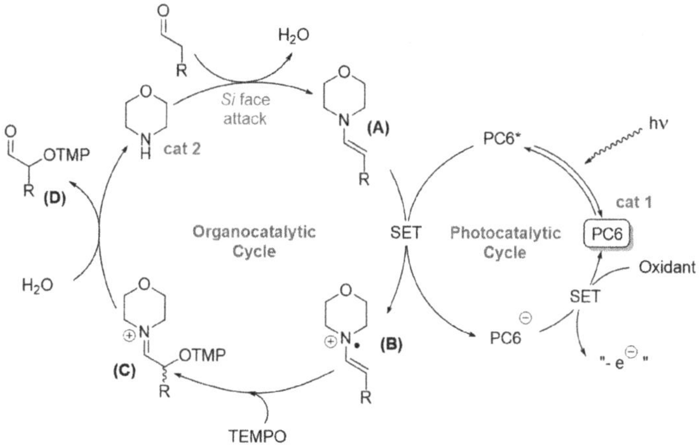

Scheme 1.57. *Photoredox organocatalytic oxyamination of aldehydes*

In 2011, Jang and co-workers reported an asymmetric version of this transformation employing semiconducting TiO_2 (Degussa P25) as photocatalyst (**PC7**), a UV light source (Hg lamp, 200–2500 nm) and several chiral secondary amines [HO 11]. While better yields were typically obtained with organocatalyst **O9**, the best enantioselectivities were reached with the fluorinated analogue **O10** (Scheme 1.58).

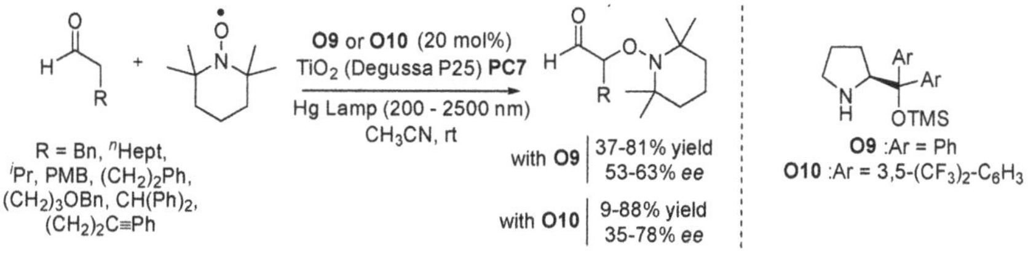

Scheme 1.58. *Asymmetric photoredox organocatalytic α-oxyamination of aldehydes*

In the same vein, Jang and co-workers described the enantioselective tandem Michael addition/α-oxyamination of enals [YOO 12].

1.3.2.1.6. α-Amination of aldehydes

In 2013, taking into account the high synthetic value of enantioenriched α-amino aldehydes, MacMillan and co-workers postulated that electrophilic carbamyl radicals would be susceptible to couple with *in situ* generated chiral enamines [CEC 13]. However, rather than employing a photoredox catalyst for promoting this transformation, the direct photoexcitation of a carbamate substrate bearing a photolabile moiety was envisioned (Scheme 1.59).

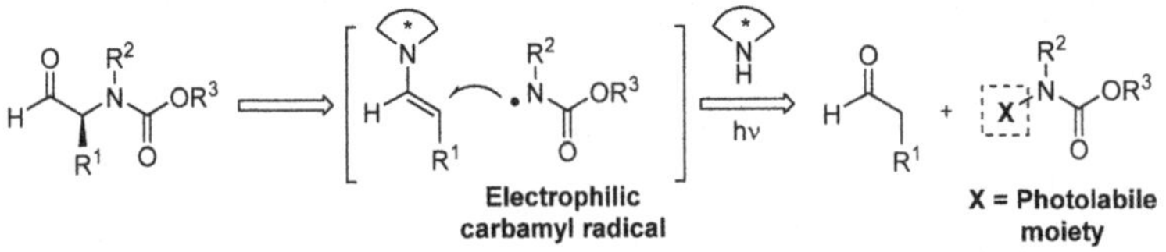

Electrophilic
carbamyl radical

X = Photolabile
moiety

Scheme 1.59. *Synthetic approach to enantioenriched α-amino aldehydes*

Seminal experiments demonstrated that the desired α-amination process could be reached when submitting hydrocinnamaldehyde to methyl 2,4-dinitrophenyl-sulfonyloxy(methyl)carbamate in the presence of an imidazolidinone organocatalyst (30 mol%) under CFL irradiation. While the authors could witness that the aminal C2 position on the imidazolidinone framework proved to be inclined to H-atom abstraction by the *N*-centered carbamyl radical, **O6** showed its poor competence such that the development of a new class of organocatalyst was required. Satisfyingly, **O15** (which possesses a quaternary C2 center) allowed the reaction to take place with improved efficiency and good stereocontrol when performed at –15°C in a DMSO/CH₃CN mixture (1:1). Under the corresponding conditions, the enantioselective α-amination of wide variety of aldehydes with a large range of 2,4-dinitrophenyl-sulfonyloxy (DNSO) carbamates was straightforwardly achieved (Scheme 1.60) [CEC 13].

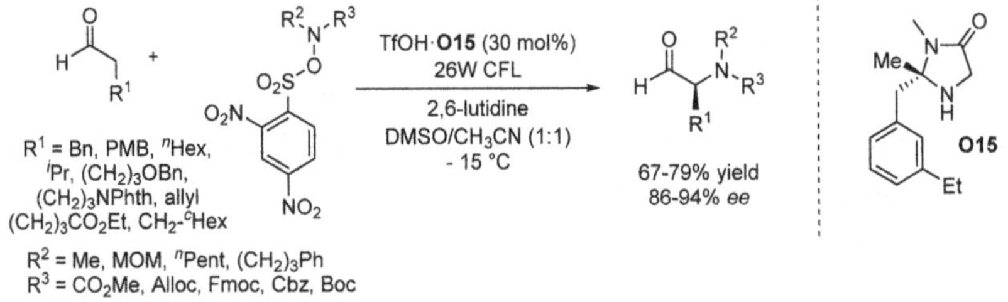

R^1 = Bn, PMB, ^{n}Hex,
^{i}Pr, (CH$_2$)$_3$OBn,
(CH$_2$)$_3$NPhth, allyl
(CH$_2$)$_3$CO$_2$Et, CH$_2$-^{c}Hex

R^2 = Me, MOM, ^{n}Pent, (CH$_2$)$_3$Ph
R^3 = CO$_2$Me, Alloc, Fmoc, Cbz, Boc

TfOH·**O15** (30 mol%)
26W CFL
2,6-lutidine
DMSO/CH₃CN (1:1)
- 15 °C

67-79% yield
86-94% *ee*

O15

Scheme 1.60. *Asymmetric photoinduced organocatalytic α-amination of aldehydes*

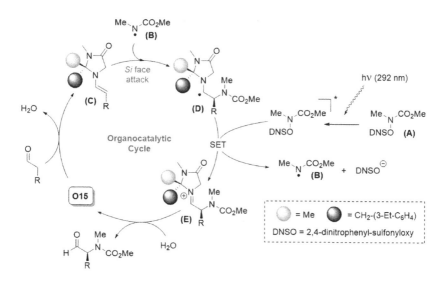

Scheme 1.61. *MacMillan's mechanistic proposal for the α-amination of aldehydes*

From a mechanistic point of view (Scheme 1.61), the authors proposed that the DNSO carbamate **A**, upon photoexcitation at 292 nm (λ_{max}(**A**)), is engaged in an SET with the α-amino radical **D** and thereby generate the carbamyl radical **B**. The latter, which possesses a significant electrophilic character, would couple with the catalytically generated nucleophilic enamine **C** and produce **D** which, involved in another SET with the excited **A***, would lead to iminium **E**. After hydrolysis, the α-aminated aldehydes would be delivered together with the regeneration of organocatalyst **O15**.

With regard to the enantioinduction effect of **O15**, DFT calculations correlated with NOESY NMR studies showed that, at the enamine stage, the more sterically hindered benzyl substituent shields the *Re* face such that the coupling with the carbamyl radical preferentially occurs on the *Si* face.

Although the authors specified that the formation an EDA complex between the transitory enamine and the DNSO carbamate could be ruled out, this mechanistic picture does not intrinsically explain how initial quantities of the carbamyl radical **B** are formed. MacMillan and co-workers further excluded the possibility of generating those through N–O bond homolysis in the absence of an SET, thus allowing us to infer that some sacrificial amounts of one reagent or intermediate does initiate the process. Even if not discussed by the authors, in view of Melchiorre's work, one eventuality would be that the photoexcited enamine promotes the required SET step [SIL 15]. Further mechanistic studies would probably clarify this point.

On the other hand, a radical chain propagation in which the α-amino radical **D** would directly reduce the fundamental DNSO carbamate **A** was excluded by the authors, since reaction progression ceased upon removal of light irradiation. However, a posterior study tends to support the likeliness of such a process [SHE 16], correlating with Yoon's conclusion that caution should be used when drawing conclusions about chain propagation from "light/dark" experiments [CIS 15]. The measurement of the quantum yield of this light-induced transformation would definitely settle the question.

1.3.2.1.7. β-Functionalization of aldehydes and ketones

In 2013, MacMillan and co-workers introduced a new enamine organocatalytic concept that, instead of permitting the traditional α-functionalization of carbonyl compounds, opened the way for the counter intuitive and valuable regioselective β-functionalization process [PIR 13]. By merging photoredox catalysis with organocatalysis, it was envisioned that a transitory enamine might lead, upon oxidative PET, to the corresponding electrophilic 3π-radical cation that, upon deprotonation of the thereby weakened β C–H bond, would induce the formation of a 5π-allylic radical. This key nucleophilic intermediate was anticipated to be susceptible to couple with synergistically photogenerated electrophilic radicals and, accordingly, produce a variety of β-functionalized carbonyl compounds in an uncharted way (Scheme 1.62).

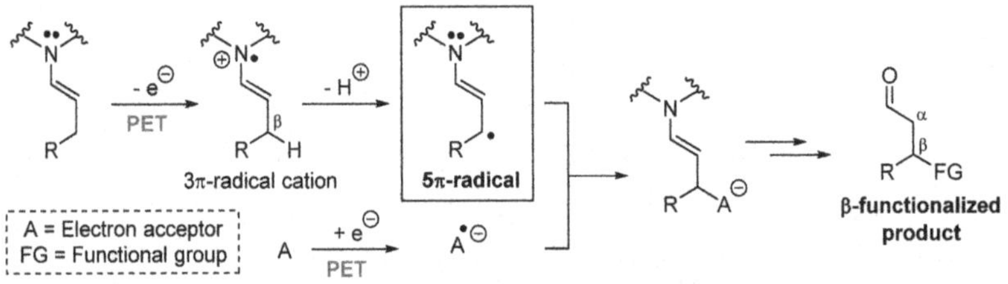

Scheme 1.62. *MacMillan's concept for the direct β-functionalization of carbonyl compounds*

The first proof of this concept was attained in the context of the direct β-arylations of aldehydes [PIR 13]. Examining the β-arylation of *n*-octanal with 1,4-dicyanobenzene, the screening a variety of amine catalysts, bases, photocatalysts and weak light sources was realized. Successfully, the desired reactivity was observed under the irradiation of a 26 CFL when the Ir(ppy)$_3$ (**PC4**, 1mol%)/*N*-isopropylbenzylamine (**O16**, 20 mol%) catalytic system was used in the presence of 1,4-diazabicyclo[2.2.2]nonane as base. Under these reaction conditions, a wide variety of aldehydes afforded the expected β-arylated products in good yields, the use of larger set of electron-deficient cyano arenes being also amenable (Scheme 1.63).

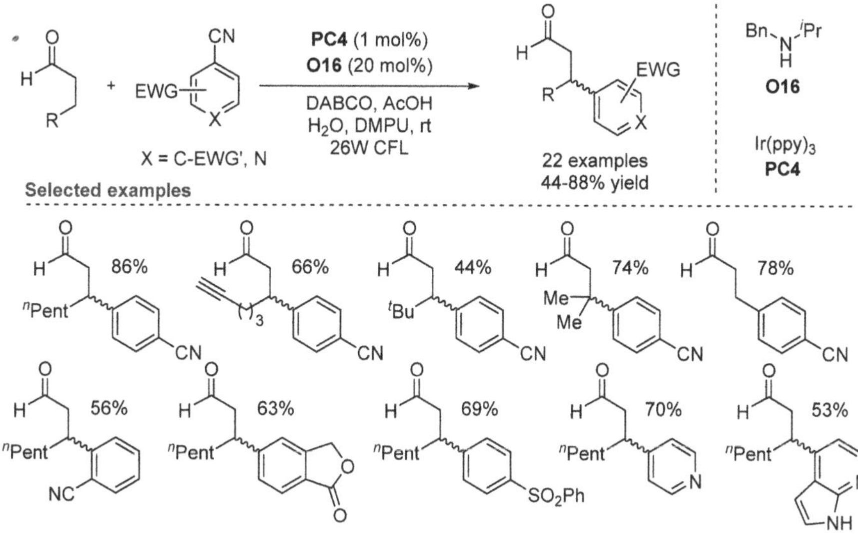

Scheme 1.63. *Direct β-functionalization of aldehydes with cyano arenes*

This work could even be extended to the use of 6-membered cyclic ketones when azepane (**O17**, 20 mol%) was used as organocatalyst. The corresponding β-aryl keto products were typically obtained in good yields, often with high diastereoselectivities (Scheme 1.64).

As a remarkable feat, using the cinchonidine-derived primary amine organocatalyst **O18**, the β-arylation of cyclohexanone with 1,4-dicaynobenzene could be performed with an encouraging level of stereoselectivity. To the best of our knowledge, up to now (spring 2017), this represents the sole asymmetric example (Scheme 1.65).

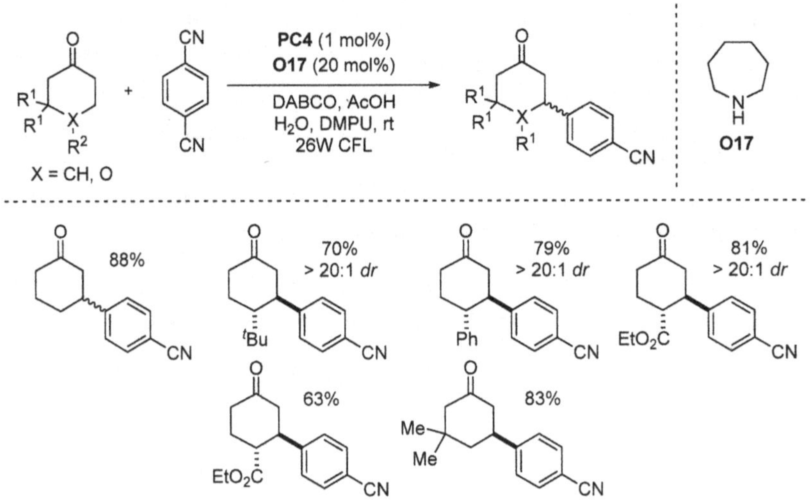

Scheme 1.64. *Direct β-functionalization of ketones with 1,4-dicyanobenzene*

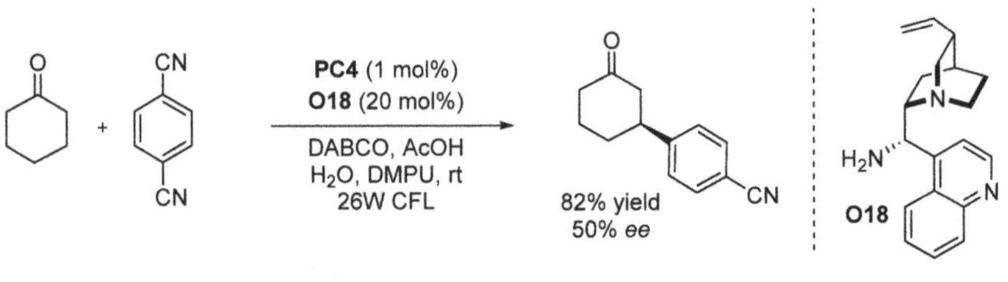

Scheme 1.65. *Enantioselective β-arylation of cyclohexanone with 1,4-dicyanobenzene*

The reaction mechanism proposed by the authors is the following. First, the PC-excited state (**PC4***), upon oxidative quench, would reduce the electron-poor aromatic compound **A** into the radical-anion **B**. In parallel, enamine **C** would be oxidized by **PC4**$^{\oplus}$ and generate the radical-cation **D** that, after deprotonation, would lead to the key 5π-electron key radical **E**. The latter would undergo a radical–radical coupling with **B** producing **F** that, after elimination of cyanide and hydrolysis, would produce the β-arylated product **G** together with the regeneration of the amine organocatalyst (Scheme 1.66).

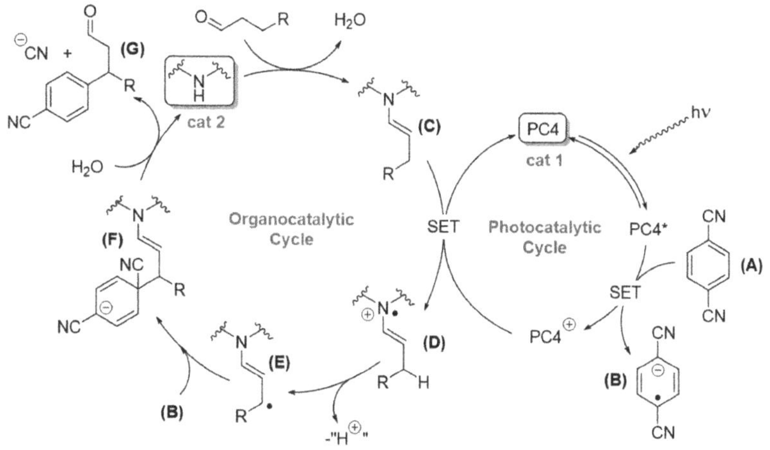

Scheme 1.66. *MacMillan's mechanistic proposal for the β-arylation of carbonyl compounds*

While we will not discuss this in more detail as this chapter is mainly focused on asymmetric transformations, it is important to know that this 5π-electron strategy has found some further developments in the racemic β-functionalization of carbonyl compounds, including alkylation's with electron-poor unsaturated alkenes and semipinacol/azapinacol-type couplings with ketones/imines [PET 13, TER 14, JEF 15].

1.3.2.2. *Iminium organocatalysis*

In contrast to enamine organocatalysis, which has known many developments in the realm of photoredox-catalyzed processes, the related iminium-based β-functionalization of a α,β-unsaturated carbonyl derivatives has long remained elusive. This in this particular context that Melchiorre and co-workers tackled this challenge with the idea that, due to the radical nature of photoredox organocatalytic processes, the enantioselective creation of an all-carbon quaternary β-stereogenic center should be practicable.

In greater detail, it was envisioned that, upon iminium ion activation of a β-substituted enone by a chiral organocatalyst, a photogenerated nucleophilic radical would effortlessly and diastereoselectively add onto the electrophilic iminium system and allow the, otherwise difficult to achieve, asymmetric radical conjugate addition process (Scheme 1.67) [MUR 16].

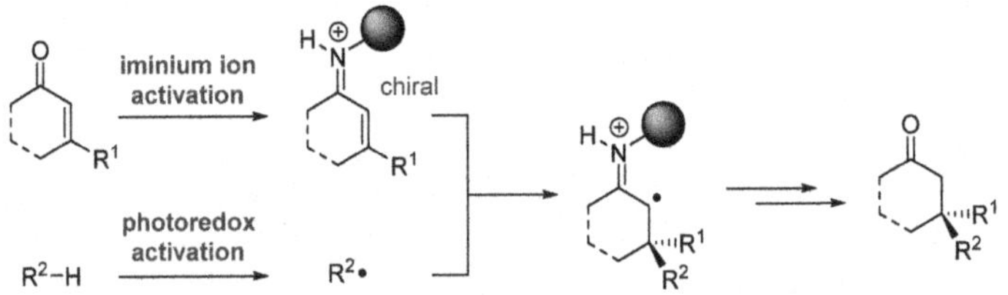

Scheme 1.67. *Melchiorre's proposition for photoredox iminium organocatalysis*

However, the success of this unprecedented strategy could not be granted from the onset. Indeed, while the coupling of the nucleophilic radical with the iminium species was expected to lead to a highly reactive α-iminyl radical-cation intermediate, the latter had already been established to be relatively unstable and predisposed to undergo a β-scission so as to reform the thermodynamically more stable iminium (Scheme 1.68) [JAK 66].

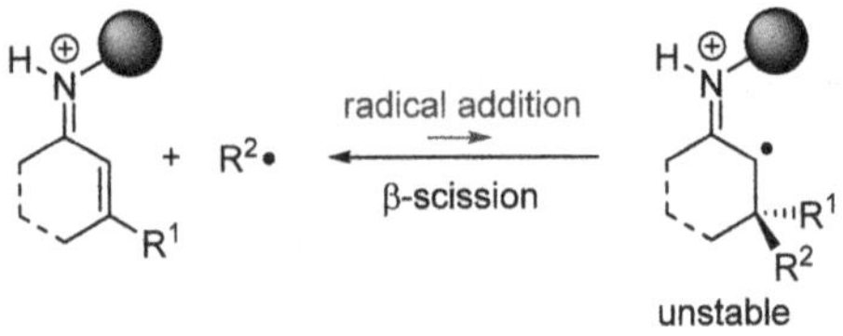

Scheme 1.68. *Stability issue associated with the radical addition step*

Mindful of this unfavorable reaction pathway, Melchiorre and co-workers ingeniously postulated that, if the organocatalyst was a primary amine possessing a redox active electron-rich moiety (electron pool), it should be possible to compete with the β-scission in favor of a fast, proximity driven SET. Thereby, a secondary enamine in rapid tautomeric equilibrium with an imine would be formed, in which hydrolysis would not only permit to deliver the expected conjugate addition product but also drive all the equilibriums toward product formation. Following this scenario, the regeneration of the organocatalyst, which entails a reduction step, was envisioned to result from a PET (Scheme 1.69).

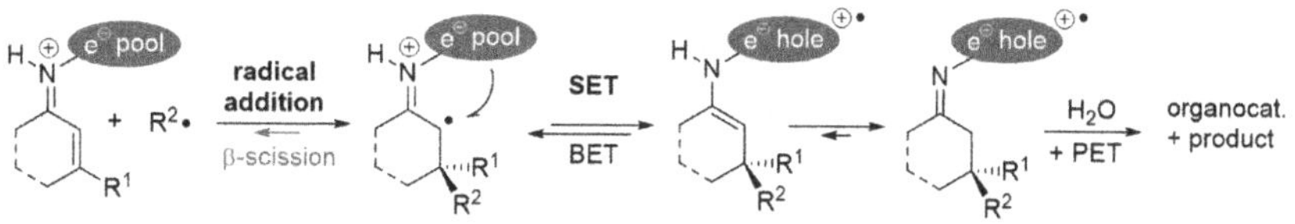

Scheme 1.69. *Melchiorre's electron pool strategy*

Taking into account all these prerequisites, the development of a new class of redox-active organocatalysts was needed. With regard to the nature of the electron pool moiety, Melchiorre *et al.* carefully chose the carbazole moiety for its excellent electron-donating capabilities and the rather high stability of the long-lived carbazole radical cation [PRU 97]. On the other hand, as a chiral linkage between the primary amine residue and the carbazole ring, the *trans*-cyclohexane scaffold was selected. Indeed, its rigidity was expected to allow reaching compact transition states and, thereby, to induce significant enantioinduction. With this careful design, the preparation of organocatalysts was undertaken **O19-O22** (Scheme 1.70).

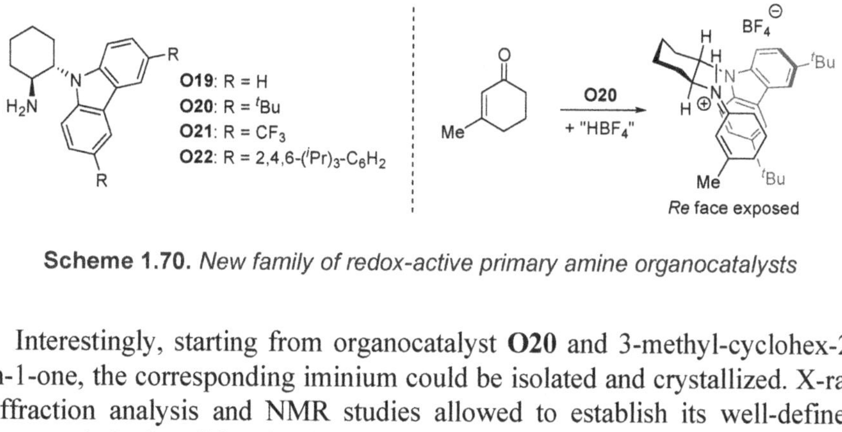

Scheme 1.70. *New family of redox-active primary amine organocatalysts*

Interestingly, starting from organocatalyst **O20** and 3-methyl-cyclohex-2-en-1-one, the corresponding iminium could be isolated and crystallized. X-ray diffraction analysis and NMR studies allowed to establish its well-defined structure in both solid and solution states (Scheme 1.65). Due to stabilizing π–π interaction between the carbazole and the iminium moieties, the C=N double bond geometry was proven to be pure (*Z*). This feature brings the carbazole in close proximity to the iminium system such that the likeliness of a stereoselective addition process (on the more exposed *Re* face) as well as that of the desired carbazole/α-iminyl radical SET could be validated.

As a first demonstration of this beautiful synthetic analysis, the enantioselective β-coupling of benzodioxoles with β-functionalized α,β-unsaturated cyclic ketones was realized. In this case, tetrabutylammonium decatungstate (**PC8**) was selected as PC by considering its recognized ability to generate benzodioxole carbon-centered radicals by cleaving one of the methylenic C–H bond via hydrogen atom transfer (HAT) [RAV 11]. With this PC (5 mol%), under the irradiation of a UV light (365 nm), the presence of both organocatalyst **O22** (20 mol%) and benzoic acid (40 mol%) in acetonitrile allowed the desired enantioselective β-functionalization of α,β-unsaturated cyclic ketones in good to quantitative yields and, considering the difficulty to asymmetrically generate all-carbon quaternary stereocenters, excellent enantioselectivities (Scheme 1.71) [MUR 16].

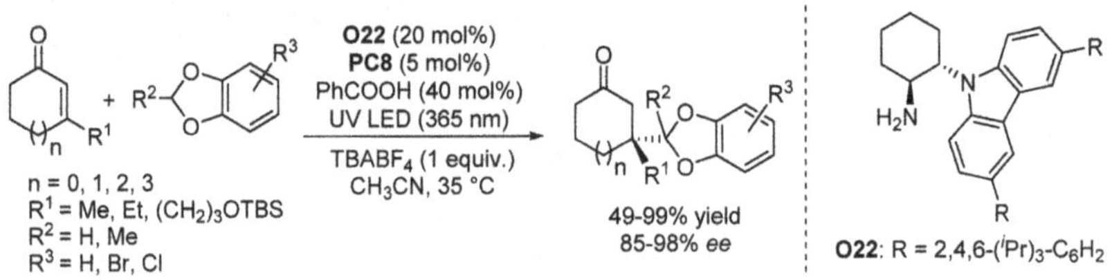

Scheme 1.71. *Enantioselective β-functionalization of cyclic enones with benzodioxoles under photoredox iminium organocatalysis*

The synthetic utility of the carbazole organocatalyst **O22** could also be expanded to the trapping of other carbon-centered radicals. Employing the commercially available PC Ir[dF(CF$_3$)ppy]$_2$(dtb-bpy)PF$_6$ (**PC9**), which is known to promote the generation of α-amino radicals from tertiary amines via oxidative SET [PRI 13], the corresponding β-aminomethylation of enones was successfully realized under the irradiation of white LEDs (λ > 400 nm) (Scheme 1.72). Interestingly, if a UV light source was used instead, this very same transformation could alternatively be performed with benzophenone as sensitizer [MUR 16].

On the basis of the above results and additional experiments, Melchiorre and co-workers proposed the following catalytic picture for these two transformations (Scheme 1.73). Upon condensation of the primary amine organocatalyst **O22** with the enone substrate, the iminium **A** would be formed. In parallel, the PC excited state, through reductive quench, would promote the generation of the nucleophilic radical **B** (by either direct HAT or SET depending on the PC used). The stereoselective radical addition of **B** onto the *Re* face of **A** would lead to the unstable α-iminyl radical **C** which, as a result of the close proximity of the organocatalyst carbazole moiety, would evolve through intramolecular SET into enamine **D** and limit, thereby, the unfavorable β-scission process. After a rapid tautomeric equilibrium into imine **E**, the previously reduced photocatalyst (PC_{red}) would give back an electron to the carbazole moiety such that, after hydrolysis, the release of the β-functionalized cyclic ketone (**G**) and the regeneration of the organocatalyst **O22** would be achieved (Scheme 1.73).

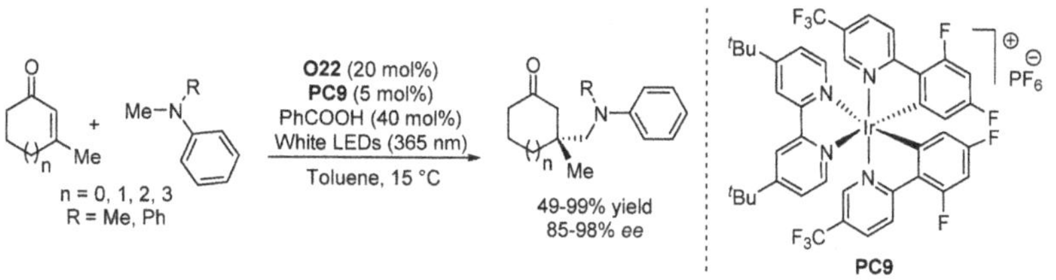

Scheme 1.72. *Enantioselective β-aminomethylation of enones with tertiary amines under photoredox iminium organocatalysis*

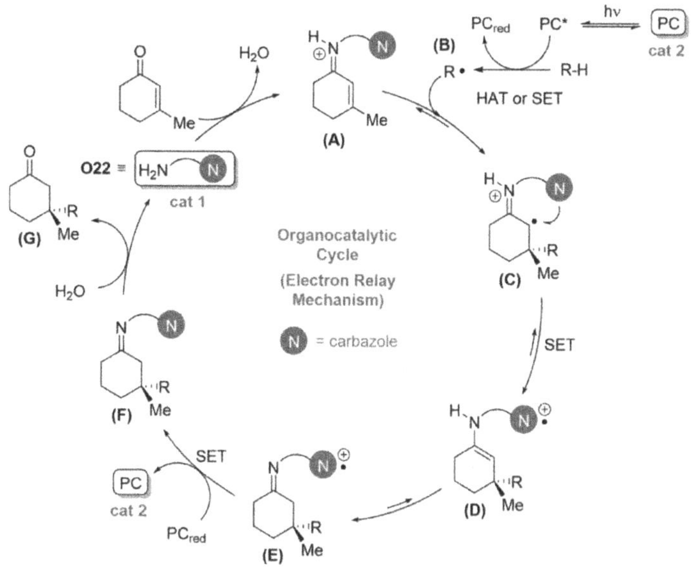

Scheme 1.73. *Mechanistic proposal for the photoredox iminium organocatalysis principle*

While Melchiorre's work has unlocked the so far uncharted field of photoredox iminium organocatalysis, it is to be expected that further developments are likely to be reached in the future.

1.3.2.3. Hydrogen bonding and Brønsted acid organocatalysis

1.3.2.3.1. Hydrogen bonding organocatalysis

As already stated, the implementation of asymmetric processes within the realm of photocatalysis has long remained a significant challenge. In this area, several research groups have investigated the possibility of taking advantage of hydrogen bonding interactions with a chiral scaffold so as to induce significant levels of enantioselectivity. While this strategy has had several successes following a stoichiometric template-based approach [BAC 99, BAC 02], the development of catalytic versions proved more complex [BRI 15].

In 2005, a major breakthrough in this field was reported by Bach and co-workers [BAU 05]. By using catalytic quantities of a benzophenone photosensitizer embedded with an amide moiety (a sort of chimera between an H-bonding organocatalyst and a PC), the authors were able to perform the intramolecular addition of an α-amino alkyl radical onto an electron-deficient C=C bond with a so far unprecedented level of enantioselectivity. Under the irradiation of a UV-lamp (λ > 300 nm), the organophotocatalyst **PC10** (30 mol%) allowed the diastereo- and enantioselective spirocyclization of a pyrrolidine-functionalized quinolone in 64% yield and 70% *ee* (Scheme 1.74) [BAU 05].

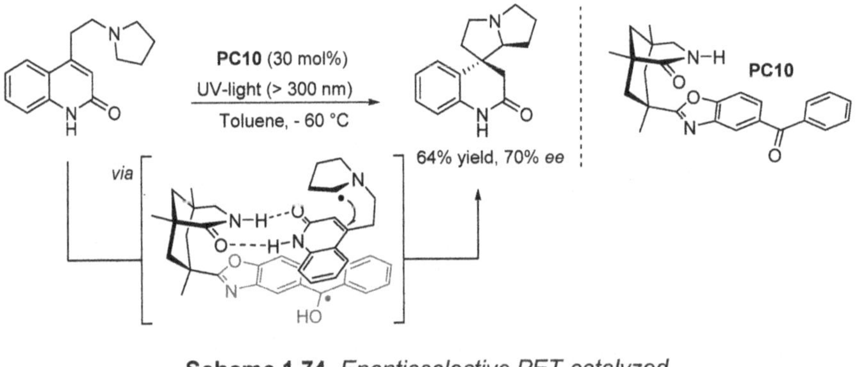

Scheme 1.74. *Enantioselective PET-catalyzed spirocyclization by hydrogen bonding*

Of particular importance, the organophotocatalyst **PC10** plays several roles in this transformation. By means of two hydrogen bonds, the substrate is brought into close proximity of the diarylketone moiety that, upon photoexcitation to the corresponding *bis*-radical ketyl species, allows the triggering of an SET with the nitrogen lone pair, followed by proton transfer (or alternatively a direct HAT). Due to the face shielding induced by the benzophenone moiety, the resulting α-amino radical is forced to attack the α,β-unsaturated system on the less hindered *Si* face (Scheme 1.74), before a back electron transfer and a protonation completes the catalytic cycle [BAU 05].

In 2009, upon refining the structure of this bifunctional organophotocatalyst, it was found that the more rigid xanthone derivative **PC11** allowed the catalytic intramolecular [2+2] cycloadditions of 4-homoallyloxy-quinolone with, at this time, an unparalleled enantioinduction (Scheme 1.75) [MUL 09].

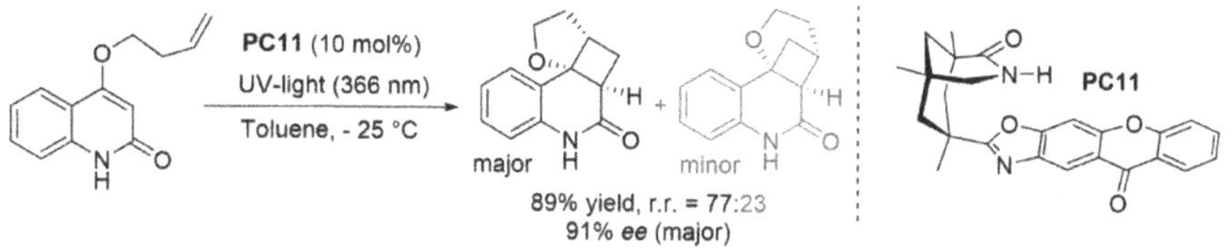

PC11 (10 mol%)
UV-light (366 nm)
Toluene, - 25 °C

major   minor
89% yield, r.r. = 77:23
91% *ee* (major)

Scheme 1.75. *Enantioselective [2+2] cycloaddition of 4-homoallyloxy-quinolone*

This enantioselective [2+2] cycloaddition process, which was proven to be applicable to a wider range of 3- and 4-subtituted quinolone derivatives [MUL 11, MAT 13], was proposed to occur through a Dexter triplet-energy transfer [DEX 53]. Hence, in this case, no SET takes places such that we cannot properly speak of photoredox organocatalysis. Yet, Bach's work further demonstrates how photochemistry and organocatalysis can synergistically be combined to reach singular radical reactivity patterns in an enantioselective fashion. Without discussing all the intricate photophysical details, a simplified mechanism is represented in Scheme 1.76.

At first, the organophotocatalyst **PC11** and the quinolone derivative, by means of hydrogen bonding interactions, would form the complex **A**. Upon photoexcitation, the xanthone residue would reach an excited triplet state that, because of its close proximity to the substrate, would allow an apparent Dexter energy transfer. Thereby, the quinolone residue would be raised to its triplet state, such that the resulting biradical species would undergo a catalyst-controlled diastereoselective radical attack onto to the pendant C=C insaturation. After a further radical–radical coupling event and dissociation of the complex, the enantioenriched [2+2] cycloaddition product would be formed and the organophotocatalyst would be regenerated (Scheme 1.76).

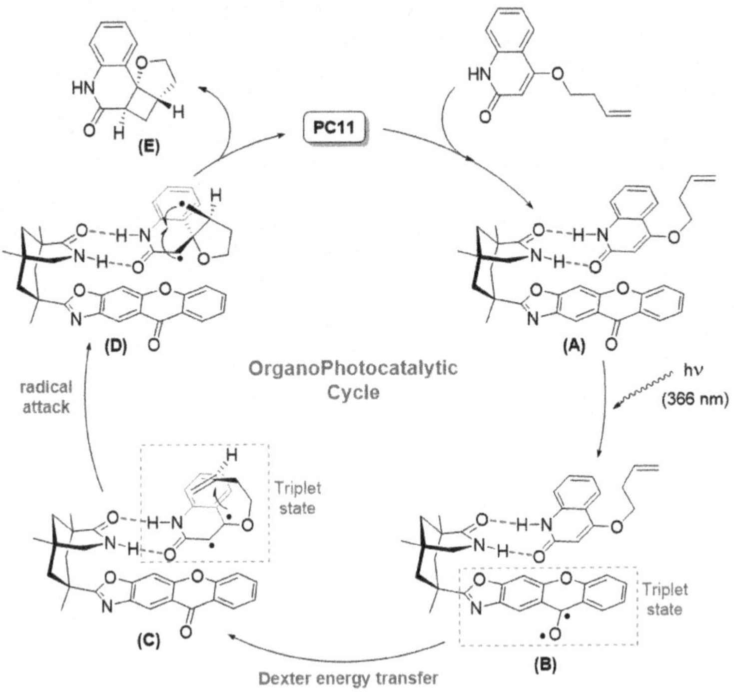

Scheme 1.76. *Enantioselective [2+2] cycloaddition of 4-homoallyloxy-quinolone*

This UVA-promoted organophotocatalytic strategy, which could successfully be applied to other enantioselective [2+2] cycloaddition processes and other classes of hydrogen-bonding catalysts [VOS 10, MAT 14, VAL 14a, VAL 14b], was later shown to be amenable to the use of visible light by replacing the xanthone moiety by a thioxanthone residue [ALO 14]. This simple modification of the catalyst structure allowed a bathochromic absorption shift of the corresponding organophotocatalyst **PC12** (λ_{max} = 387 nm) such that the previously reported enantioselective [2+2] cycloaddition of quinolone could be performed conveniently under visible light irradiation, often with better yields and enantioselectivities. Due to the enhanced stability of **PC12**, which is less prone to undergo decomposition via HAT upon photoexcitation, the akin intermolecular enantioselective [2+2] photocycloaddition of quinolones with α,β-unsaturated carbonyl compounds could even be performed (Scheme 1.77) [TRO 16].

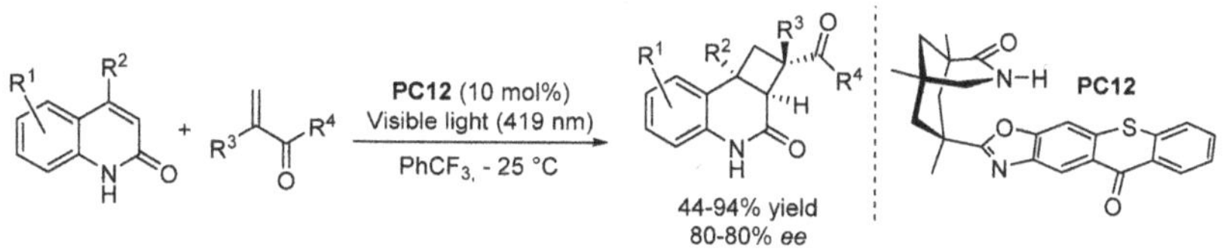

Scheme 1.77. *Visible light mediated enantioselective intermolecular [2+2] cycloadditions*

Notably, in 2016 Melchiorre and co-workers reported the cooperative association of photocatalysis with hydrogen bonding organocatalysis for the enantioselective functionalization of photochemically generated hydroxy-*o*-quinodimethanes [DEL 16, HEP 16]. Due to the polar nature of the organocatalytic cycle, we will not discuss these stimulating results herein.

1.3.2.3.2. Brønsted acid organocatalysis

In 2016, Melchiorre and co-workers reported the Brønsted acid catalyzed and photoredox-catalyzed addition of α-amino radicals onto alkenylpyridines. When submitting a variety of 2-alkenylpyridines to various tertiary aromatic amines in the presence of diphenylphosphoric acid (**O23**, 5 mol%) and $Ir[dF(CF_3)ppy]_2(dtb-bpy)PF_6$ (**PC9**, 1 mol%) under the irradiation of blue LEDs (465 nm), the corresponding β-amino derivatives were obtained in good yields (Scheme 1.78). Importantly, the presence of both Brønsted acid and the PC was required for this reaction to efficiently take place, suggesting the synergistic nature of this catalytic process.

With enantioselectivity that might allow room for improvement, this reaction could notably be performed in an asymmetric fashion by replacing **O23** with the chiral BINOL-derived phosphoric acid **O24** (Scheme 1.78). The catalytic cycle of this asymmetric photoredox organocatalytic transformation is represented hereafter.

The authors suggest that, upon reductive quench, the PC-excited state (**PC9***) would promote the generation of the requisite nucleophilic α-amino radical **A** (via oxidation followed by a deprotonation) from the corresponding tertiary aromatic amine. On the other hand, the alkenylpyridine would be protonated by the Brønsted acid organocatalyst and form the electrophilic chiral ion pair **B**. Diastereoselective radical addition of **A** onto **B** would induce the formation of the enantioenriched α-pyridinium radical **C**, inclined to be subsequently reduced into **D** via the concomitant regeneration of the PC. After tautomeric equilibrium, the enantioenriched product **E** would be released while a protonation event would reintroduce the Brønsted acid organocatalyst into the catalytic cycle (Scheme 1.79).

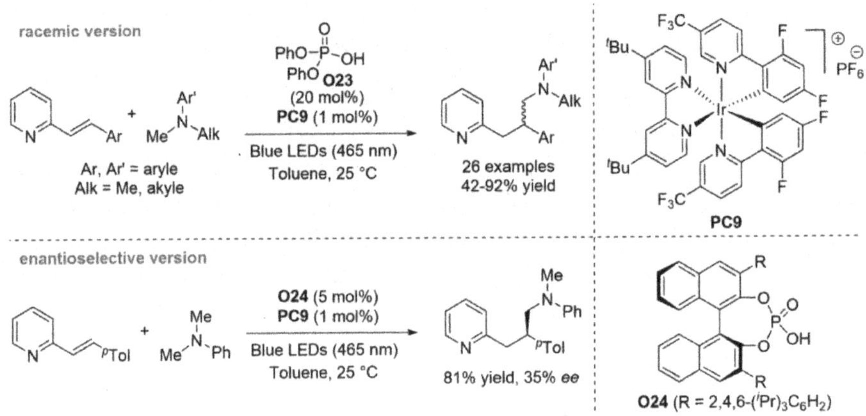

Scheme 1.78. *Synergistic merger of Brønsted acid catalysis and photoredox catalysis for the β-aminomethylation of alkenylpyridines*

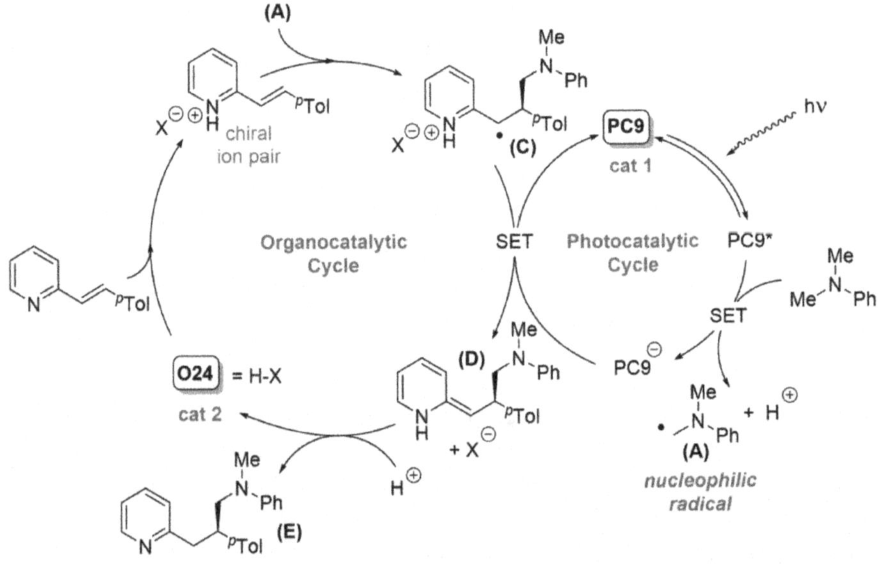

Scheme 1.79. *Mechanistic proposal for the β-aminomethylation of alkenylpyridines*

In 2013, resulting from their long interest in proton-coupled electron transfer (PCET) initiated processes [HOF 17, YAY 14, GEN 16], Knowles and co-workers reported the development of an asymmetric aza-pinacol cyclization reaction by merging the use of a chiral Brønsted acid organocatalyst with that of a photoredox catalyst [RON 13]. Submitting various hydrazino-ketones to Ir(ppy)$_2$(dtb-bpy)PF$_6$ (**PC5**, 2 mol%) and the chiral phosphoric acid **O25** (10 mol%) in the presence of Hantzsch dihydropyridine (HEH, 1.5 equiv.) under the irradiation of blue LEDs, the corresponding cyclic 1,2-amino alcohols were typically obtained with perfect diastereoselectivity, good yields and remarkable enantioselectivities (Scheme 1.80).

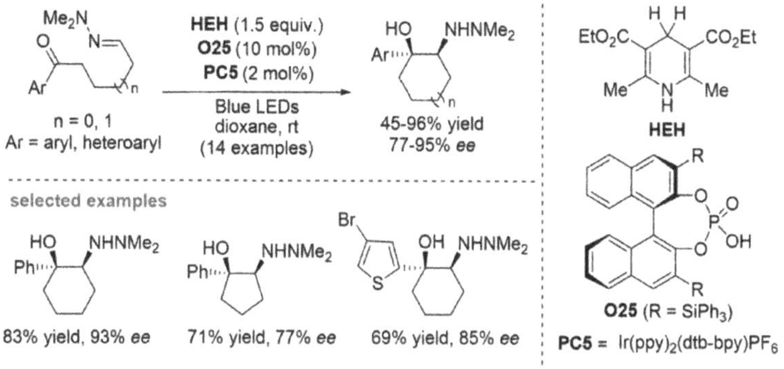

Scheme 1.80. *PCET-enabled asymmetric aza-pinacol cyclization*

Continuing from their previous catalytic generation of ketyl radical by concerted reduction/protonation (PCET) [TAR 13], the authors envisioned that the ketone moiety, upon PCET with the chiral phosphoric acid and the iridium PC, could lead to a neutral ketyl intermediate susceptible to preserving a hydrogen bonding interaction with the chiral phosphate. Accordingly, it was anticipated that, if this interaction could be maintained on the time scale of the subsequent rapid aza-pinacol coupling, the chiral phosphate could transfer its stereogenic information such that the cyclization process would operate in an asymmetric manner.

The development of the desired enantioselective aza-pinacol cyclization met the expected success, suggesting the validity of the following catalytic cycle (Scheme 1.81).

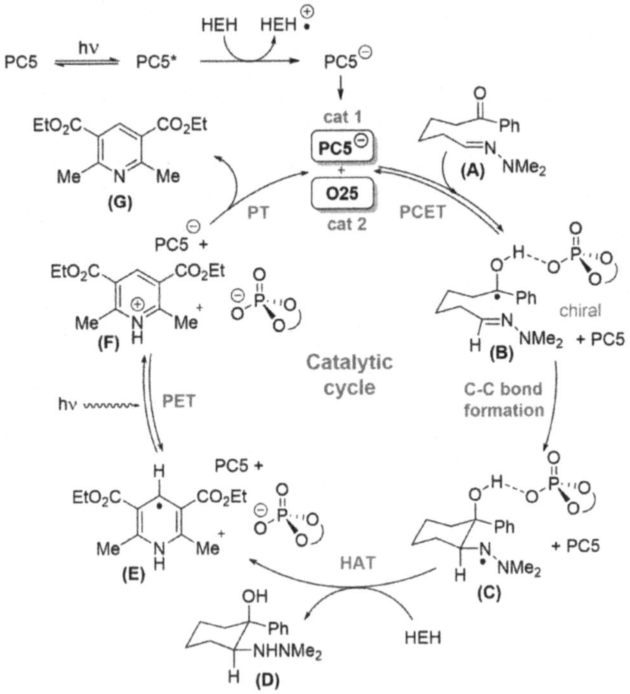

Scheme 1.81. *Mechanistic proposal for the PCET-enabled asymmetric aza-pinacol cyclization*

At first, upon sacrificial oxidation of HEH, the excited PC would generate the first catalyst, **PC5**$^{\ominus}$. The latter, in synergy with the Brønsted acid organocatalyst **O25**, would then induce a concerted reduction/protonation of the C=O bond of **A** leading to the chiral phosphate bounded ketyl radical **B**. Upon diastereoselective C–C bond formation, the resulting hydrazyl radical **C** would be engaged in a hydrogen atom transfer with the Hantzsch's ester (HEH) affording the desired enantioenriched cyclized product **D** associated with the formation of the corresponding HEH radical **E**. Finally, after PET and proton transfer, both catalysts would be regenerated with the release of pyridine **G**.

By the end of 2016, to the best of our knowledge, this work represents the first example of PCET-mediated asymmetric reaction and opens many perspective for the development of new synergistic photoredox organocatalytic transformations based on the use of chiral Brønsted acids.

1.3.2.4. *Ion-pairing organocatalysis*

Ion-pairing organocatalysis, which encompasses phase transfer catalysis (PTC), is a powerful organocatalytic strategy [BRI 12, BRA 13]. Naturally, its synergistic association with photochemical processes has also been investigated. Since 2012, numerous efforts have been devoted to the α-hydroxylation of β-keto carbonyl coupounds via the photochemical sensitization of molecular oxygen into singlet oxygen [LIA 12, WAN 16a, WAN 16b]. However, the first examples in which this synergistic association was employed in radical-based C–C bond formations were independently disclosed by Melchiorre's and Ooi's groups in 2015 [WOZ 15, URA 15].

Based on their previous work on the PC-free EDA complex activation concept, Melchiorre and co-workers questioned whether it would be possible to realize the enantioselective perfluoroalkylation of β-ketoesters by relying on an *in situ* generated chiral ion pair. At the onset, it was envisioned that an enolate should be a sufficient electron-donor so as to form a CT complex with a perfluoroalkyl iodide which, upon photoexcitation, could lead to the generation of valuable electrophilic perfluoroalkyl radicals. In the presence of a chiral phase transfer catalyst, it was envisioned that the enolate partner would exist under the form of a chiral ion-pair and, thereby, couple with such electron-deficient radicals in an asymmetric fashion.

This strategy proved successful and, while the formation of the desired CT complexes could be evidenced, the perfluoroalkylation of a wide range of indanone-derived β-ketoesters could be performed enantioselectively under the irradiation of white LEDs in the presence of the cinchonine-derived organocatalyst **O26** (20 mol%) [WOZ 15]. Employing caesium carbonate as base (2 equiv.) and a chlorobenzene/perfluorooctane solvent mixture, the desired products were typically obtained in average to good yields and excellent asymmetric inductions, better results being obtained when the indanone aromatic ring was bearing electron-withdrawing substituents (Scheme 1.82).

From a mechanistic perspective, the authors proposed a radical chain propagation pathway. Upon deprotonation of the starting β-keto ester by phase transfer catalysis with **O26**, the chiral ion pair **A** is suggested to form the EDA complex **B** with the perfluoroalkyl iodide. Upon its photoexcitation, catalytic quantities of perfluoroalkyl radical **C** would be formed (radical initiation), susceptible to coupling diastereoselectively with the chiral ion pair **A**. The ketyl intermediate **D** would then undergo an iodide atom transfer with the starting electrophilic iodide, thereby promoting radical chain propagation, while **E** would rapidly collapse and regenerate the PTC and liberate the fluoroalkyl product **F** (Scheme 1.83).

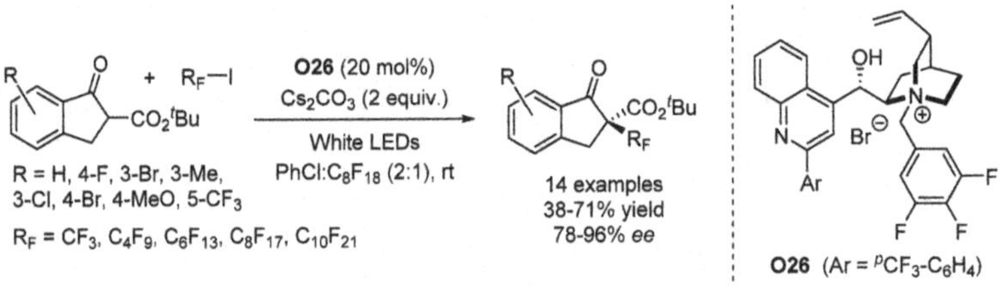

Scheme 1.82. *Photo-organocatalytic enantioselective perfluoroalkylation of β-ketoesters*

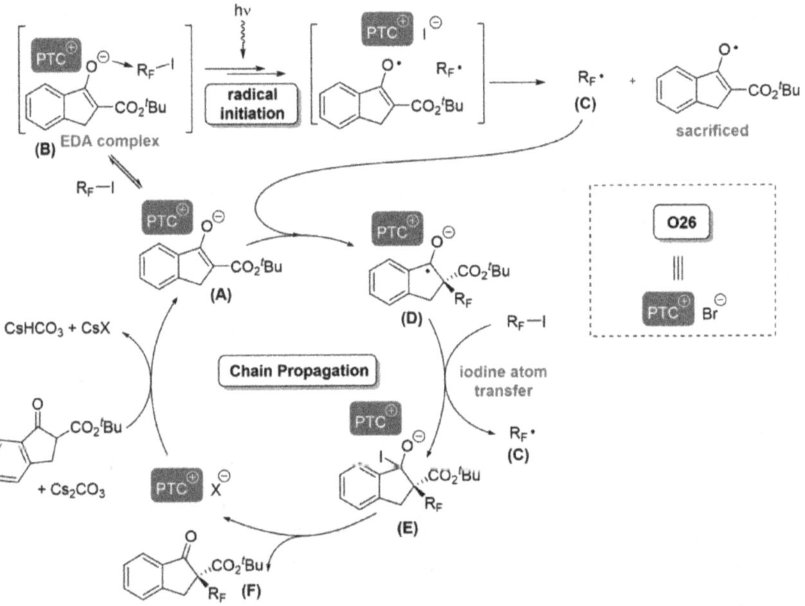

Scheme 1.83. *Mechanistic proposal for Melchiorre's perfluoroalkylation reaction*

Later the same year, based on a similar strategy in which photocatalysis is merged with chiral ion-pair catalysis, Ooi and co-workers reported that, under visible light irradiation, the tetraaminophosphonium salt **O27**

(4 mol%) and photoredox catalyst **PC13** (1 mol%) promoted the redox-neutral asymmetric α-coupling of *N*-arylaminomethanes with *N*-sulfonyl aldimines. Following this strategy, a wide variety of enantioenriched 1,2-diamines could be obtained with good to excellent yields and enantioselectivities (Scheme 1.84) [URA 15].

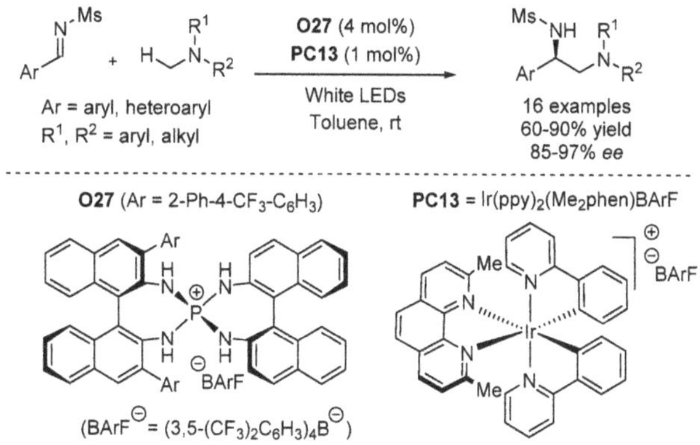

Scheme 1.84. *Ooi's asymmetric α-coupling of N-arylaminomethanes with aldimines*

In this transformation, the authors proposed that the excited photocatalyst **PC13***, upon reductive quench, permits both the formation of an α-aminoalkyl radical **A** (by oxidation of *N*-methyl arylamines) and that of an α-aminobenzyl radical anion **B** (by reduction of the imine partner). While the latter is believed to form the chiral ion-pair **C** with **O27**, the radical–radical coupling event is thought to occur stereoselectively because of the intimate proximity of the chiral cation (Scheme 1.85) [URA 15].

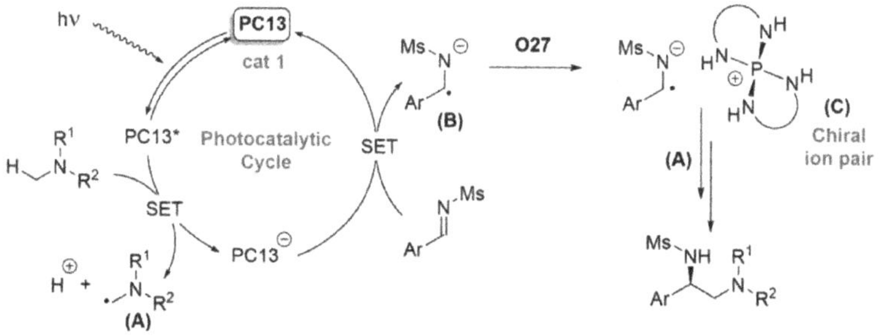

Scheme 1.85. *Ooi's mechanistic proposal for the enantioselective α-coupling of N-arylaminomethanes with aldimines*

While other reaction mechanisms are not totally excluded by Ooi and co-workers, this transformation was later shown to occur under a reversed redox manner by capitalizing on the oxidative quench of Ir(ppy)$_3$* (**PC4***), provided that α-trimethylsilyl N-methyl arylamines are used [KIZ 16].

1.4. Conclusions

In conclusion, we hope that the readers are now convinced of the extraordinary outcomes that could be reached by merging organocatalysis with radical processes. Indeed, while it represents a unique opportunity to harness the high reactivity of radical intermediates in an asymmetric fashion, it opens the way for the development of otherwise unachievable catalytic processes. From initial developments in which oxidative enamine SOMO-organocatalysis made it possible to perform original umpolung-based α-functionalizations of carbonyl derivatives, the advent of photocatalyzed synergistic processes allowed the development of a fantastic variety of (pseudo) redox-neutral transformations.

This particular organocatalytic strategy, which is without any doubt in its infancy, may be expected to reach even more stimulating developments in the future. Although not part of this review, recent developments showed that NHC enantioselective organocatalysis could also be combined with the use of single electron oxidants [WHI 14, ZHA 15, WHI 15]. Therefore, it is conceivable that the current state of the art could soon find other applications within the realm of other organocatalytic methods and that *"radical organocatalysis"* might be a chance for the discovery of innovative activation ganomodes [HAS 14]. Not only have all these advents had a significant impact on the synthetic utility of organocatalysis but, in a more general way, they have reintroduced radical-based processes as key technologies for the development of enantioselective transformations.

1.5. Bibliography

[AFE 16] AFEWERKI S., CÓRDOVA A., "Combinations of aminocatalysts and metal catalysts: a powerful cooperative approach in selective organic synthesis", *Chemical Reviews*, vol. 116, no. 22, pp. 13512–13570, 2016.

[ALO 14] ALONSO R., BACH T., "A chiral thioxanthone as an organocatalyst for enantioselective [2+2] photocycloaddition reactions induced by visible light", *Angewandte Chemie International Edition*, vol. 53, no. 17, pp. 4368–4371, 2014.

[AMA 09] AMATORE M., BEESON T.D., BROWN S.P. *et al.*, Enantioselective linchpin catalysis by SOMO catalysis: an approach to the asymmetric α-chlorination of aldehydes and terminal epoxide formation", *Angewandte Chemie International Edition*, vol. 48, no. 28, pp. 5121–5124, 2009.

[ARC 13] ARCEO E., JURBERG I. D., ÁLVAREZ-FERNÁNDEZ A. *et al.*, "Photochemical activity of a key donor–acceptor complex can drive stereoselective catalytic α-alkylation of aldehydes", *Nature Chemistry*, vol. 5, no. 9, pp. 750–756, 2013.

[ARC 14] ARCEO E., BAHAMONDE A., BERGONZINI G. *et al.*, Enantioselective direct α-alkylation of cyclic ketones by means of photo-organocatalysis", *Chemical Science*, vol. 5, no. 6, pp. 2438–2442, 2014.

[BAC 99] BACH T., BERGMANN H., HARMS K., "High facial diastereoselectivity in the photocycloaddition of a chiral aromatic aldehyde and an enamide induced by intermolecular hydrogen bonding", *Journal of the American Chemical Society*, vol. 121, no. 45, pp. 10650–10651, 1999.

[BAC 02] BACH T., BERGMANN H., GROSCH B. *et al.*, "Highly enantioselective intra- and intermolecular [2+2] photocycloaddition reactions of 2-quinolones mediated by a chiral lactam host: host-guest interactions, product configuration, and the origin of the stereoselectivity in solution", *Journal of the American Chemical Society*, vol. 124, no. 27, pp. 7982–7990, 2002.

[BAH 16] BAHAMONDE A., MELCHIORRE P., "Mechanism of the stereoselective α-alkylation of aldehydes driven by the photochemical activity of enamines", *Journal of the American Chemical Society*, vol. 138, no. 25, pp. 8019–8030, 2016.

[BAL 15] BALZANI V., BERGAMINI G., CERONI P., "Light: a very peculiar reactant and product", *Angewandte Chemie International Edition*, vol. 54, no. 39, pp. 11320–11337, 2015.

[BAU 05] BAUER A., WESTKÄMPER F., GRIMME S. *et al.*, "Catalytic enantioselective reactions driven by photoinduced electron transfer", *Nature*, vol. 436, no. 7054, pp. 1139–1140, 2005.

[BEE 11] BEEL R., KOBIALKA S., SCHMIDT M.L. *et al.*, "Direct experimental evidence for an enamine radical cation in SOMO catalysis", *Chemical Communications*, vol. 47, no. 11, pp. 3293–3295, 2011.

[BEE 07] BEESON T.D., MASTRACCHIO A., HONG J.-B. *et al.*, "Enantioselective organocatalysis using SOMO activation", *Science*, vol. 316, no. 5824, pp. 582–585, 2007.

[BER 79] VAN BERGEN T.J., HEDSTRAND D. M., KRUIZINGA W.H. *et al.*, "Chemistry of dihydropyridines. 9. Hydride transfer from 1,4-dihydropyridines to sp³-hybridized carbon in sulfonium salts and activated halides. Studies with NAD(P)H models", *Journal of Organic Chemistry*, vol. 44, no. 26, pp. 4953–4962, 1979.

[BER 14] BERGONZINI G., SCHINDLER C.S., WALLENTIN C.J. *et al.*, "Photoredox activation and anion binding catalysis in the dual catalytic enantioselective synthesis of β-amino esters", *Chemical Science*, vol. 5, no. 1, pp. 112–116, 2014.

[BOC 79] BOCK C.R., CONNOR J.A., GUTIERREZ A.R. *et al.*, "Estimation of excited-state redox potentials by electron-transfer quenching. Applications of electron-transfer theory to excited-state redox processes", *Journal of the American Chemical Society*, vol. 101, no. 17, pp. 4815–4824, 1979.

[BRA 13] BRAK K., JACOBSEN E.N., "Asymmetric ion-pairing catalysis", *Angewandte Chemie International Edition*, vol. 52, no. 2, pp. 534–561, 2013.

[BRI 12] BRIÈRE J.F., OUDEYER S., DALLA V. *et al.*, "Recent advances in cooperative ion pairing in asymmetric organocatalysis", *Chemical Society Reviews*, vol. 41, no. 5, pp. 1696–1707, 2012.

[BRI 15] BRIMIOULLE R., LENHART D., MATURI M. *et al.*, "Enantioselective catalysis of photochemical reactions", *Angewandte Chemie International Edition*, vol. 54, no. 13, pp. 3872–3890, 2015.

[BRO 04] BROCHU M.P., BROWN S.P., MACMILLAN D.W.C., "Direct and enantioselective organocatalytic α-chlorination of aldehydes", *Journal of the American Chemical Society*, vol. 126, no. 13, pp. 4108–4109, 2004.

[BRO 10] BROWN A.R., KUO W.H., JACOBSEN E.N., "Enantioselective catalytic α-alkylation of aldehydes via an S_N1 pathway", *Journal of the American Chemical Society*, vol. 132, no. 27, pp. 9286–9288, 2010.

[BUI 09] BUI N.N., HO X.H., MHO S.I.L. *et al.*, "Organocatalyzed α-oxyamination of aldehydes using anodic oxidation", *European Journal of Organic Chemistry*, vol. 2009, no. 31, pp. 5309–5312, 2009.

[CAN 10] CANNIZZO A., MILNE C.J., CONSANI C. *et al.*, "Light-induced spin crossover in Fe(II)-based complexes: the full photocycle unraveled by ultrafast optical and X-ray spectroscopies", *Coordination Chemistry Reviews*, vol. 254, no. 21–22, pp. 2677–2686, 2010.

[CAN 84a] CANO-YELO H., DERONZIER A., "Photo-oxidation of some carbinols by the Ru(II) polypyridyl complex-aryl diazonium salt system", *Tetrahedron Letters*, vol. 25, no. 48, pp. 5517–5520, 1984.

[CAN 84b] CANO-YELO H., DERONZIER A., "Photocatalysis of the Pschorr reaction by tris-(2,2'-bipyridyl)ruthenium(II) in the phenanthrene series", *Journal of the Chemical Society, Perkin Transactions 2*, no. 6, pp. 1093–1098, 1984.

[CAN 75] CANTACUZENE D., DORME R., "Cetones α perfluorées", *Tetrahedron Letters*, vol. 16, no. 25, pp. 2031–2034, 1975.

[CAN 77] CANTACUZÈNE D., WAKSELMAN C., DORME R., "Condensation of perfluoroalkyl iodides with unsaturated nitrogen compounds", *Journal of the Chemical Society, Perkin Transactions 1*, no. 12, pp. 1365–1371, 1977.

[CAS 65] CASTRO C.E., GAUGHAN E.J., OWSLEY D.C., "Cupric halide halogenations", *Journal of Organic Chemistry*, vol. 30, no. 2, pp. 587–592, 1965.

[CEC 13] CECERE G., KÖNIG C.M., ALLEVA J.L. *et al.*, "Enantioselective direct α-amination of aldehydes via a photoredox mechanism: a strategy for asymmetric amine fragment coupling", *Journal of the American Chemical Society*, vol. 135, no. 31, pp. 11521–11524, 2013.

[CHE 12] CHEREVATSKAYA M., NEUMANN M., FÜLDNER S. *et al.*, "Visible-light-promoted stereoselective alkylation by combining heterogeneous photocatalysis with organocatalysis", *Angewandte Chemie International Edition*, vol. 51, no. 17, pp. 4062–4066, 2012.

[CIS 15] CISMESIA M.A., YOON T.P., "Characterizing chain processes in visible light photoredox catalysis", *Chemical Science*, vol. 6, no. 10, pp. 5426–5434, 2015.

[COM 13] COMITO R.J., FINELLI F.G., MACMILLAN D.W.C., "Enantioselective intramolecular aldehyde α-alkylation with simple olefins: direct access to homo-ene products", *Journal of the American Chemical Society*, vol. 135, no. 25, pp. 9358–9361, 2013.

[CON 09] CONRAD J.C., KONG J., LAFORTEZA B.N. *et al.*, "Enantioselective α-arylation of aldehydes via organo-SOMO catalysis. An ortho-selective arylation reaction based on an open-shell pathway", *Journal of the American Chemical Society*, vol. 131, no. 33, pp. 11640–11641, 2009.

[CÓR 04] CÓRDOVA A., SUNDÉN H., ENGQVIST M. *et al.*, "The direct amino acid-catalyzed asymmetric incorporation of molecular oxygen to organic compounds", *Journal of the American Chemical Society*, vol. 126, no. 29, pp. 8914–8915, 2004.

[COR 16] CORRIGAN N., SHANMUGAM S., BOYER C., "Photocatalysis in organic and polymer synthesis", *Chemical Society Reviews*, vol. 36, 2016, pp. 385–394, 2016.

[COS 93a] COSSY J., BOUZIDE A., "Radical cyclization of *N*-alkyl-*N*-unsaturated alkyl-β-carboxamidoenamines induced by manganese acetate", *Synlett*, no. 3, pp. 202–204, 1993.

[COS 93b] COSSY J., BOUZIDE A., "Generation of radical cations from enamines and their addition to unactivated olefins", *Journal of the Chemical Society, Chemical Communications*, no. 15, pp. 1218–1219, 1993.

[DEL 16] DELLAMICO L., VEGA-PEÑALOZA A., CUADROS S. *et al.*, "Enantioselective organocatalytic Diels-Alder trapping of photochemically generated hydroxy-*o*-quinodimethanes", *Angewandte Chemie International Edition*, vol. 55, no. 10, pp. 3313–3317, 2016.

[DEV 10] DEVERY J.J., CONRAD J.C., MACMILLAN D.W.C. *et al.*, "Mechanistic complexity in organo-SOMO activation", *Angewandte Chemie International Edition*, vol. 49, no. 35, pp. 6106–6110, 2010.

[DEX 53] DEXTER D.L., "A theory of sensitized luminescence in solids", *Journal of Chemical Physics*, vol. 21, no. 5, pp. 836–850, 1953.

[DHI 01] DHIMANE A.-L., FENSTERBANK L., MALACRIA M., "Polycyclic compounds via radical cascade reactions", in RENAUD P., SIBI M.P. (eds), *Radicals in Organic Synthesis*, Wiley, Weinheim, 2001.

[END 08] ENDERS D., WANG C., BATS J.W., "Organocatalytic asymmetric domino reactions: a cascade consisting of a Michael addition and an aldehyde α-alkylation", *Angewandte Chemie International Edition*, vol. 47, no. 39, pp. 7539–7542, 2008.

[FEN 14] FENG Z.J., XUAN J., XIA X.D. *et al.*, "Direct sp^3 C–H acroleination of *N*-aryl-tetrahydroisoquinolines by merging photoredox catalysis with nucleophilic catalysis", *Organic & Biomolecular Chemistry*, vol. 12, no. 13, pp. 2037, 2014.

[FID 12] FIDALY K., CEBALLOS C., FALGUIERES A. *et al.*, "Visible light photoredox organocatalysis: a fully transition metal-free direct asymmetric α alkylation of aldehydes", *Green Chemistry*, vol. 14, no. 5, pp. 1293–1297, 2012.

[FOS 80] FOSTER R., "Electron donor-acceptor complexes", *Journal of Physical Chemistry*, vol. 84, no. 17, pp. 2135–2141, 1980.

[GEN 16] GENTRY E.C., KNOWLES R.R., "Synthetic applications of proton-coupled electron transfer", *Accounts of Chemical Research*, vol. 49, no. 8, pp. 1546–1556, 2016.

[GRA 08] GRAHAM T.H., JONES C.M., JUI N.T. *et al.*, "Enantioselective organo-singly occupied molecular orbital catalysis: the carbo-oxidation of styrenes", *Journal of the American Chemical Society*, vol. 130, no. 49, pp. 16494–16495, 2008.

[GUA 15] GUALANDI A., MARCHINI M., MENGOZZI L. *et al.*, "Organocatalytic enantioselective alkylation of aldehydes with [Fe(bpy)₃]Br₂ catalyst and visible light", *ACS Catalysis*, vol. 5, no. 10, pp. 5927–5931, 2015.

[HAS 14] HASHIMOTO T., KAWAMATA Y., MARUOKA K., "An organic thiyl radical catalyst for enantioselective cyclization", *Nature Chemistry*, vol. 6, pp. 702–705, 2014.

[HED 78] HEDSTRAND D.M., KRUIZINGA W.H., KELLOGG R.M., "Light induced and dye accelerated reductions of phenacyl onium salts by 1,4-dihydropyridines", *Tetrahedron Letters*, vol. 19, no. 14, pp. 1255–1258, 1978.

[HEP 16] HEPBURN H., MAGAGNANO G., MELCHIORRE P., "Light-triggered enantioselective organocatalytic Mannich-type reaction" *Synthesis*, vol. 49, no. 1, pp. 76–86, 2016.

[HO 10] HO X.H., MHO S.I.L, KANG H. *et al.*, "Electro-organocatalysis: enantioselective α-alkylation of aldehydes", *European Journal of Organic Chemistry*, vol. 132, pp. 4436–4441, 2010.

[HO 11] HO X.H., KANG M.J., KIM S.J. *et al.*, "Green organophotocatalysis. TiO₂-induced enantioselective α-oxyamination of aldehydes", *Catalysis Science & Technology*, vol. 1, no. 6, pp. 923–926, 2011.

[HOF 17] HOFFMANN N., "Proton-coupled electron transfer in photoredox catalytic reactions", *European Journal of Organic Chemistry*, no. 15, pp. 1982–1992, 2017.

[IBR 06] IBRAHEM I., ZHAO G.L., SUNDÉN H. *et al.*, "A route to 1,2-diols by enantioselective organocatalytic α-oxidation with molecular oxygen", *Tetrahedron Letters*, vol. 47, no. 27, pp. 4659–4663, 2006.

[ISM 13] ISMAILI H., PITRE S.P., SCAIANO J.C., "Active participation of amine-derived radicals in photoredox catalysis as exemplified by a reductive cyclization", *Catalysis Science & Technology*, vol. 3, no. 4, pp. 935–937, 2013.

[JAK 66] JAKOBSEN H.J., LAWESSON S.O., MARSHALL J.T.B. *et al.*, "Studies in mass spectrometry. Part XII. Mass spectra of enamines", *Journal of the Chemical Society B*, pp. 940–946. 1966.

[JAN 07] JANG H.Y., HONG J.B., MACMILLAN D.W.C., "Enantioselective organocatalytic singly occupied molecular orbital activation: the enantioselective α-enolation of aldehydes", *Journal of the American Chemical Society*, vol. 129, no. 22, pp. 7004–7005, 2007.

[JEF 15] JEFFREY J.L., PETRONIJEVIĆ F.R., MACMILLAN D.W.C., Selective radical–radical cross-couplings: design of a formal β-Mannich reaction", *Journal of the American Chemical Society*, vol. 137, no. 26, pp. 8404–8407, 2015.

[JUI 10] JUI N.T., LEE E.C.Y., MACMILLAN D.W.C., "Enantioselective organo-SOMO cascade cycloadditions: a rapid approach to molecular complexity from simple aldehydes and olefins", *Journal of the American Chemical Society*, vol. 132, no. 29, pp. 10015–10017, 2010.

[JUI 12] JUI N.T., GARBER J.A.O., FINELLI F.G. *et al.*, "Enantioselective organo-SOMO cycloadditions: a catalytic approach to complex pyrrolidines from olefins and aldehydes", *Journal of the American Chemical Society*, vol. 134, no. 28, pp. 11400–11403, 2012.

[KIM 08] KIM H., MACMILLAN D.W.C., "Enantioselective organo-SOMO catalysis: the α-vinylation of aldehydes", *Journal of the American Chemical Society*, vol. 130, no. 2, pp. 398–399, 2008.

[KIZ 16] KIZU T., URAGUCHI D., OOI T., "Independence from the sequence of single-electron transfer of photoredox process in redox-neutral asymmetric bond-forming reaction", *Journal of Organic Chemistry*, vol. 81, no. 16, pp. 6953–6958, 2016.

[KOI 09] KOIKE T., AKITA M., "Photoinduced oxyamination of enamines and aldehydes with TEMPO catalyzed by [Ru(bpy)$_3$]$^{2+}$", *Chemistry Letters*, vol. 38, no. 2, pp. 166–167, 2009.

[KOI 14] KOIKE T., AKITA M., "Visible-light radical reaction designed by Ru- and Ir-based photoredox catalysis", *Inorganic Chemistry Frontiers*, vol. 1, no. 8, pp. 562–576, 2014.

[LAK 12] LAKHDAR S., MAJI B., MAYR H., "Imidazolidinone-derived enamines: nucleophiles with low reactivity", *Angewandte Chemie International Edition*, vol. 51, no. 23, pp. 5739–5742, 2012.

[LAN 16] LANG X., ZHAO, J., CHEN X., "Cooperative photoredox catalysis", *Chemical Society Reviews*, vol. 45, pp. 3026–3038, 2016.

[LIA 12] LIAN M., LI Z., CAI Y. *et al.*, "Enantioselective photooxygenation of β-keto esters by chiral phase-transfer catalysis using molecular oxygen", *Chemistry Asian Journal*, vol. 7, no. 9, pp. 2019–2023, 2012.

[LIM 16] LIMA C.G.S., LIMA T.D.M., DUARTE M. *et al.*, "Organic synthesis enabled by light-irradiation of EDA complexes: theoretical background and synthetic applications", *ACS Catalysis*, vol. 6, no. 3, pp. 1389–1407, 2016.

[LIS 14] LIST B., ĆORIĆ I., GRYGORENKO O.O. *et al.*, "The catalytic asymmetric α-benzylation of aldehydes", *Angewandte Chemie International Edition*, vol. 53, no. 1, pp. 282–285, 2014.

[MAS 10] MASTRACCHIO A., WARKENTIN A.A., WALJI A.M. *et al.*, "Direct and enantioselective α-allylation of ketones via singly occupied molecular orbital (SOMO) catalysis", *Proceedings of the National Academy of Sciences of the United States of America*, vol. 107, no. 48, pp. 20648–20651, 2010.

[MAT 13] MATURI M.M., WENNINGER M., ALONSO R. *et al.*, "Intramolecular [2+2] photocycloaddition of 3- and 4-(but-3-enyl) oxyquinolones: influence of the alkene substitution pattern, photophysical studies, and enantioselective catalysis by a chiral sensitizer", *Chemistry A European Journal*, vol. 19, no. 23, pp. 7461–7472, 2013.

[MAT 14] MATURI M.M., BACH T., "Enantioselective catalysis of the intermolecular [2+2] photocycloaddition between 2-pyridones and acetylenedicarboxylates", *Angewandte Chemie International Edition*, vol. 53, no. 29, pp. 7661–7664, 2014.

[MAY 12] MAYR H., LAKHDAR S., MAJI B. *et al.*, "A quantitative approach to nucleophilic organocatalysis", *Beilstein Journal of Organic Chemistry*, vol. 8, pp. 1458–1478, 2012.

[MEG 15] MEGGERS E., "Asymmetric catalysis activated by visible light", *Chemical Communications*, vol. 51, no. 16, pp. 3290–3301, 2015.

[MOR 13] MORI T., INOUE Y., "Charge-transfer excitation: unconventional yet practical means for controlling stereoselectivity in asymmetric photoreactions", *Chemical Society Reviews*, vol. 42, no. 20, pp. 8122–8133, 2013.

[MUL 09] MÜLLER C., BAUER A., BACH T., "Light-driven enantioselective organocatalysis", *Angewandte Chemie International Edition*, vol. 48, no. 36, pp. 6640–6642, 2009.

[MUL 11] MÜLLER C., BAUER A., MATURI M.M. *et al.*, "Enantioselective intramolecular [2+2]-photocycloaddition reactions of 4-substituted quinolones catalyzed by a chiral sensitizer with a hydrogen-bonding motif", *Journal of the American Chemical Society*, vol. 133, no. 41, pp. 16689–16697, 2011.

[MUL 50] MULLIKEN R.S., "Structures of complexes formed by halogen molecules with aromatic and with oxygenated solvents 1", *Journal of the American Chemical Society*, vol. 72, no. 1, pp. 600–608, 1950.

[MUL 52] MULLIKEN R.S., "Molecular compounds and their spectra. II", *Journal of the American Chemical Society*, vol. 74, no. 8, pp. 811–824, 1952.

[MUR 16] MURPHY J.J., BASTIDA D., PARIA S. *et al.*, "Asymmetric catalytic formation of quaternary carbons by iminium ion trapping of radicals", *Nature*, vol. 532, no. 7598, pp. 218–222, 2016.

[NAG 09] NAGIB D.A., SCOTT M.E., MACMILLAN D.W.C., "Enantioselective α-trifluoromethylation of aldehydes via photoredox organocatalysis", *Journal of the American Chemical Society*, vol. 131, no. 31, pp. 10875–10877, 2009.

[NAR 92] NARASAKA K., OKAUCHI T., TANAKA K. *et al.*, "Generation of cation radicals from enamines and their reactions with olefins", *Chemistry Letters*, vol. 21, no. 10, pp. 2099–2102, 1992.

[NEU 11] NEUMANN M., FÜLDNER S., KÖNIG B. *et al.*, "Metal-free, cooperative asymmetric organophotoredox catalysis with visible light", *Angewandte Chemie International Edition*, vol. 50, no. 4, pp. 951–954, 2011.

[NEU 12] NEUMANN M., ZEITLER K., "Application of microflow conditions to visible light photoredox catalysis", *Organic Letters*, vol. 14, no. 11, pp. 2658–2661, 2012.

[NIC 08] NICEWICZ D.A., MACMILLAN D.W.C., "Merging photoredox catalysis with organocatalysis: the direct asymmetric alkylation of aldehydes", *Science*, vol. 322, no. 5898, pp. 77–80, 2008.

[NIC 09] NICOLAOU K.C., REINGRUBER R., SARLAH D. *et al.*, "Enantioselective intramolecular Friedel-Crafts-type α-arylation of aldehydes", *Journal of the American Chemical Society*, vol. 131, no. 6, pp. 2086–2087, 2009.

[OIS 78] OISHI S., FURUTA N., "Quenching of the luminescent excited state of tris(2,2'-bipyridyl)ruthenium(II) with bromonitromethane", *Chemistry Letters*, vol. 7, no. 1, pp. 45–48, 1978.

[PET 13] PETRONIJEVIĆ F.R., NAPPI M., MACMILLAN D.W.C., "Direct β-functionalization of cyclic ketones with aryl ketones via the merger of photoredox and organocatalysis", *Journal of the American Chemical Society*, vol. 135, no. 49, pp. 18323–18326, 2013.

[PHA 11] PHAM P.V., ASHTON K., MACMILLAN D.W.C., "The intramolecular asymmetric allylation of aldehydes via organo-SOMO catalysis: a novel approach to ring construction", *Chemical Science*, vol. 2, no. 8, pp. 1470–1473, 2011.

[PIR 13] PIRNOT M.T., RANKIC D.A., MARTIN D.B.C. *et al.*, "Photoredox activation for the direct β-arylation of ketones and aldehydes", *Science*, vol. 339, no. 6127, pp. 1593–1596, 2013.

[PIV 86] PIVA O., HENIN F., MUZART J. *et al.*, "Enantioselective photodeconjugation of α,β-unsaturated esters: effect of the nature of the chiral agent", *Tetrahedron Letters*, vol. 27, no. 26, pp. 3001–3004, 1986.

[PIV 90] PIVA O., MORTEZAEI R., HENIN F. *et al.*, "Highly enantioselective photodeconjugation of α,β.-unsaturated esters. Origin of the chiral discrimination", *Journal of the American Chemical Society*, vol. 112, no. 25, pp. 9263–9272, 1990.

[PRI 13] PRIER C.K., RANKIC D.A., MACMILLAN D.W.C., "Visible light photoredox catalysis with transition metal complexes: applications in organic synthesis", *Chemical Reviews*, vol. 113, no. 7, pp. 5322–5363, 2013.

[PRU 97] PRUDHOMME D.R., WANG Z., RIZZO C.J., "An improved photosensitizer for the photoinduced electron-transfer deoxygenation of benzoates and *m*-(trifluoromethyl)benzoates", *Journal of Organic Chemistry*, vol. 62, no. 23, pp. 8257–8260, 1997.

[RAV 11] RAVELLI D., ALBINI A., FAGNONI M., "Smooth photocatalytic preparation of 2-substituted 1,3-benzodioxoles", *Chemistry A European Journal*, vol. 17, no. 2, pp. 572–579, 2011.

[REN 90] RENAUD P., SCHUBERT S., "Stereoselective addition of carbon-centered radicals to chiral enamines", *Synlett*, vol. 1990, no. 10, pp. 624–626, 1990.

[REN 10] RENDLER S., MACMILLAN D.W.C., "Enantioselective polyene cyclization via organo-SOMO catalysis", *Journal of the American Chemical Society*, vol. 132, no. 14, pp. 5027–5029, 2010.

[RIE 14] RIENTE P., MATAS ADAMS A., ALBERO J. *et al.*, "Light-driven organocatalysis using inexpensive, nontoxic Bi_2O_3 as the photocatalyst", *Angewandte Chemie International Edition*, vol. 53, no. 36, pp. 9613–9616, 2014.

[ROM 16] ROMERO N.A., NICEWICZ D.A., "Organic photoredox catalysis", *Chemical Reviews*, vol. 116, no. 17, pp. 10075–10166, 2016.

[RON 13] RONO L.J., YAYLA H.G., WANG D.Y. *et al.*, "Enantioselective photoredox catalysis enabled by proton-coupled electron transfer: development of an asymmetric aza-pinacol cyclization", *Journal of the American Chemical Society*, vol. 135, no. 47, pp. 17735–17738, 2013.

[ROT 16] ROTH H.G., ROMERO N.A., NICEWICZ D.A., "Experimental and calculated electrochemical potentials of common organic molecules for applications to single-electron redox chemistry", *Synlett*, vol. 27, no. 5, pp. 714–723, 2016.

[RUS 91] RUSSELL G.A., WANG K., "El+ectron transfer processes. 53. Homolytic alkylation of enamines by electrophilic radicals", *Journal of Organic Chemistry*, vol. 56, no. 11, pp. 3475–3479, 1991.

[SHA 16] SHAW M.H., TWILTON J., MACMILLAN D.W.C., "Photoredox catalysis in organic chemistry", *Journal of Organic Chemistry*, vol. 81, no. 16, pp. 6898–6926, 2016.

[SHE 16] SHEN X., HARMS K., MARSCH M. *et al.*, "A rhodium catalyst superior to iridium congeners for enantioselective radical amination activated by visible light", *Chemistry A European Journal*, vol. 22, no. 27, pp. 9102–9105, 2016.

[SHI 10] SHIH H.W., VANDER WAL M.N., GRANGE R.L. *et al.*, "Enantioselective α-benzylation of aldehydes via photoredox organocatalysis", *Journal of the American Chemical Society*, vol. 132, no. 39, pp. 13600–13603, 2010.

[SIB 07] SIBI M.P., HASEGAWA M., "Organocatalysis in radical chemistry. Enantioselective α-oxyamination of aldehydes", *Journal of the American Chemical Society*, vol. 129, no. 14, pp. 4124–4125, 2007.

[SIL 15] SILVI M., ARCEO E., JURBERG I.D. *et al.*, "Enantioselective organocatalytic alkylation of aldehydes and enals driven by the direct photoexcitation of enamines", *Journal of the American Chemical Society*, vol. 137, no. 19, pp. 6120–6123, 2015.

[SIM 12] SIMONOVICH S.P., VAN HUMBECK J.F., MACMILLAN D.W.C., "A general approach to the enantioselective α-oxidation of aldehydes via synergistic catalysis", *Chemical Science*, vol. 3, pp. 58–61, 2012.

[SKU 16] SKUBI K.L., BLUM T.R., YOON T.P., "Dual catalysis strategies in photochemical synthesis", *Chemical Reviews*, vol. 116, no. 17, pp. 10035–10074, 2016.

[SUN 04] SUNDÉN H., ENGQVIST M., CASAS J. *et al.*, "Direct amino acid catalyzed asymmetric α oxidation of ketones with molecular oxygen", *Angewandte Chemie International Edition*, vol. 43, no. 47, pp. 6532–6535, 2004.

[TAR 13] TARANTINO K.T., LIU P., KNOWLES R.R., "Catalytic ketyl-olefin cyclizations enabled by proton-coupled electron transfer", *Journal of the American Chemical Society*, vol. 135, no. 27, pp. 10022–10025, 2013.

[TER 14] TERRETT J.A., CLIFT M.D., MACMILLAN D.W.C., "Direct β-alkylation of aldehydes via photoredox organocatalysis", *Journal of the American Chemical Society*, vol. 136, no. 19, pp. 6858–6861, 2014.

[TIS 14] TISOVSKÝ P., MEČIAROVÁ M., ŠEBESTA R., "Asymmetric organocatalytic SOMO reactions of enol silanes and silyl ketene (thio)acetals", *Organic & Biomolecular Chemistry*, vol. 12, no. 46, pp. 9446–9452, 2014.

[TOB 72] TOBINAGA S., KOTANI E., "Intramolecular and intermolecular oxidative coupling reactions by a new iron complex [Fe(DMF)$_3$Cl$_2$][FeCl$_4$]", *Journal of the American Chemical Society*, vol. 94, no. 1, pp. 309–310, 1972.

[TOT 16] TÓTH B.L., TISCHLER O., NOVÁK Z., "Recent advances in dual transition metal–visible light photoredox catalysis", *Tetrahedron Letters*, vol. 57, no. 41, pp. 4505–4513, 2016.

[TRO 16] TRÖSTER A., ALONSO R., BAUER A. *et al.*, "Enantioselective intermolecular [2+2] photocycloaddition reactions of 2(1*H*)-quinolones induced by visible light irradiation", *Journal of the American Chemical Society*, vol. 138, no. 25, pp. 7808–7811, 2016.

[UM 10] UM J.M., GUTIERREZ O., SCHOENEBECK F. *et al.*, "Nature of intermediates in organo-somo catalysis of α-arylation of aldehydes", *Journal of the American Chemical Society*, vol. 132, no. 17, pp. 6001–6005, 2010.

[URA 15] URAGUCHI D., KINOSHITA N., KIZU T. *et al.*, "Synergistic catalysis of ionic brønsted acid and photosensitizer for a redox neutral asymmetric α-coupling of N-arylaminomethanes with aldimines", *Journal of the American Chemical Society*, vol. 137, no. 43, pp. 13768–13771, 2015.

[VAN 10] VAN HUMBECK J.F., SIMONOVICH S.P., KNOWLES R.R. *et al.*, "Concerning the mechanism of the FeCl₃-catalyzed α-oxyamination of aldehydes: evidence for a non-SOMO activation pathway", *Journal of the American Chemical Society*, vol. 132, no. 29, pp. 10012–10014, 2010.

[VAL 14a] VALLAVOJU N., SELVAKUMAR S., JOCKUSCH S. *et al.*, "Enantioselective organo-photocatalysis mediated by atropisomeric thiourea derivatives", *Angewandte Chemie International Edition*, vol. 53, no. 22, pp. 5604–5608, 2014.

[VAL 14b] VALLAVOJU N., SELVAKUMAR S., JOCKUSCH S. *et al.*, "Evaluating thiourea architecture for intramolecular [2+2] photocycloaddition of 4-alkenylcoumarins", *Advanced Synthesis & Catalysis*, vol. 356, no. 13, pp. 2763–2768, 2014.

[VIG 04] VIGNOLA N., LIST B., "Catalytic asymmetric intramolecular α-alkylation of aldehydes", *Journal of the American Chemical Society*, vol. 126, no. 2, pp. 450–451, 2004.

[VOS 10] VOSS F., BACH T., "An ethynyl-substituted 1,5,7-trimethyl-3-azabicyclo[3.3.1]nonan-2-one as a versatile precursor for chiral templates and chiral photocatalysts", *Synlett*, vol. 2010, no. 10, pp. 1493–1496, 2010.

[WAL 15] WALASZEK D.J., RYBICKA-JASIŃSKA K., SMOLEŃ S. *et al.*, "Mechanistic insights into enantioselective C-H photooxygenation of aldehydes via enamine catalysis", *Advanced Synthesis & Catalysis*, vol. 357, no. 9, pp. 2061–2070, 2015.

[WAN 15] WANG C., LU Z., "Catalytic enantioselective organic transformations via visible light photocatalysis", *Organic Chemistry Frontiers*, vol. 2, no. 2, pp. 179–190, 2015.

[WAN 16a] WANG Y., ZHENG Z., LIAN M. *et al.*, "Photo-organocatalytic enantioselective α-hydroxylation of β-keto esters and β-keto amides with oxygen under phase transfer catalysis", *Green Chemistry*, vol. 18, no. 20, pp. 5493–5499, 2016.

[WAN 16b] WANG Y., YIN H., TANG X. *et al.*, "A series of cinchona-derived *N*-oxide phase-transfer catalysts: application to the photo-organocatalytic enantioselective α-hydroxylation of β-dicarbonyl compounds", *Journal of Organic Chemistry*, vol. 81, no. 16, pp. 7042–7050, 2016.

[WEI 16] WEI G., ZHANG C., BUREŠ F. *et al.*, "Enantioselective aerobic oxidative $C(sp^3)$–H olefination of amines via cooperative photoredox and asymmetric catalysis", *ACS Catalysis*, vol. 6, no. 6, pp. 3708–3712, 2016.

[WEL 15] WELIN E.R., WARKENTIN A.A., CONRAD J.C. *et al.*, "Enantioselective α-alkylation of aldehydes by photoredox organocatalysis: rapid access to pharmacophore fragments from β-cyanoaldehydes", *Angewandte Chemie International Edition*, vol. 54, no. 33, pp. 9668–9672, 2015.

[WHI 14] WHITE N.A., ROVIS T., "Enantioselective N-heterocyclic carbene-catalyzed β-hydroxylation of enals using nitroarenes: an atom transfer reaction that proceeds via single electron transfer", *Journal of the American Chemical Society*, vol. 136, no. 42, pp. 14674–14677, 2014.

[WHI 15] WHITE N.A., ROVIS T., "Oxidatively initiated NHC-catalyzed enantioselective synthesis of 3,4-disubstituted cyclopentanones from enals", *Journal of the American Chemical Society*, vol. 137, no. 32, pp. 10112–10115, 2015.

[WIL 09] WILSON J.E., CASAREZ A.D., MACMILLAN D.W.C., "Enantioselective aldehyde α-nitroalkylation via oxidative organocatalysis", *Journal of the American Chemical Society*, vol. 131, no. 32, pp. 11332–11334, 2009.

[WOZ 15] WOŹNIAK Ł., MURPHY J.J., MELCHIORRE P., "Photo-organocatalytic enantioselective perfluoroalkylation of β-ketoesters", *Journal of the American Chemical Society*, vol. 137, no. 17, pp. 5678–5681, 2015.

[WU 12] WU P., HE C., WANG J. *et al.*, "Photoactive chiral metal-organic frameworks for light-driven asymmetric α-alkylation of aldehydes", *Journal of the American Chemical Society*, vol. 134, no. 36, pp. 14991–14999, 2012.

[XU 14] XU C., ZHANG L., LUO S., "Merging aerobic oxidation and enamine catalysis in the asymmetric α-amination of β-ketocarbonyls using N-hydroxycarbamates as nitrogen sources", *Angewandte Chemie International Edition*, vol. 53, no. 16, pp. 4149–4153, 2014.

[YAN 16] YANG M., YANG X., SUN H. *et al.*, "Total synthesis of ileabethoxazole, pseudopteroxazole, and seco-pseudopteroxazole", *Angewandte Chemie International Edition*, vol. 55, no. 8, pp. 2851–2855, 2016.

[YAY 14] YAYLA H., KNOWLES R., "Proton-coupled electron transfer in organic synthesis: novel homolytic bond activations and catalytic asymmetric reactions with free radicals", *Synlett*, vol. 25, no. 20, pp. 2819–2826, 2014.

[YOD 05] YODER R.A., JOHNSTON J.N., "A case study in biomimetic total synthesis: polyolefin carbocyclizations to terpenes and steroids", *Chemical Reviews*, vol. 105, no. 12, pp. 4730–4756, 2005.

[YOO 12] YOON H.S., HO X.H., JANG J. *et al.*, "N719 dye-sensitized organophotocatalysis: enantioselective tandem michael addition/oxyamination of aldehydes", *Organic Letters*, vol. 14, no. 13, pp. 3272–3275, 2012.

[ZHA 15] ZHANG Y., DU Y., HUANG Z. *et al.*, "*N*-Heterocyclic carbene-catalyzed radical reactions for highly enantioselective β-hydroxylation of enals", *Journal of the American Chemical Society*, vol. 137, no. 7, pp. 2416–2419, 2015.

[ZHA 16] ZHANG X., "Mechanism and enantioselectivity in α-alkylation of carbonyl compounds via photoredox organocatalysis: a DFT study", *Computational and Theoretical Chemistry*, vol. 1078, pp. 113–122, 2016.

[ZHU 14] ZHU Y., ZHANG L., LUO S., "Asymmetric α-photoalkylation of β-ketocarbonyls by primary amine catalysis: facile access to acyclic all-carbon quaternary stereocenters", *Journal of the American Chemical Society*, vol. 136, no. 42, pp. 14642–14645, 2014.

[ZIV 16] ZIVIC N., BOUZRATI-ZERELLI M., KERMAGORET A. *et al.*, "Photocatalysts in polymerization reactions", *ChemCatChem*, vol. 8, no. 9, pp. 1617–1631, 2016.

2

Chiral Quaternary Ammonium
Salts in Organocatalysis

2.1. Introduction

The acceleration of a reaction by means of a substoichiometric amount of small chiral organic molecules emerged as a powerful asymmetric synthetic methodology during the early 2000s [BER 05, DAL 13]. As testified by the increasing number of publications, *organocatalysis* has become the third pillar of catalysis beside metallocatalysis and biocatalysis. In a modern society looking for both efficient and environmentally benign processes, catalysis, especially when making use of usually environmentally benign organic-based catalysts, stands out when talking of a contemporary sustainable chemistry. The term *organocatalysis* was coined in 2000 [AHR 00], along with the conceptualization of generic modes of activation allowing the classification of organocatalytic processes [BER 05, MAC 08, SEA 05]. Nevertheless, the occurrences of organic catalysts in synthesis dated back much before this period of time [MAC 08]. Quaternary ammonium salts ($R_4N^+X^-$) is a representative example of this nonlinear historical evolution. In the late 1960s [MAR 08b, SHI 13b, NEL 99, OOI 07b, HAS 07, JEW 09, SHI 13A, HER 14], leading works by Starks, Makosza and Brändström established that amphiphilic quaternary phosphonium or ammonium salts, having long alkyl chains R, were able to accelerate a reaction occurring in a biphasic mixture, either solid/liquid or liquid/liquid immiscible phases (Scheme 2.1.(a)). In 1971, Starks proposed the term "phase-transfer catalysis" (PTC) to account for the ability of R_4N^+

Chapter written by Sylvain OUDEYER, Vincent LEVACHER and Jean-François BRIÈRE.

cation (i) to accelerate the transport a polar nucleophile Nu^- species from one polar phase (a solid phase such as a mineral base or an aqueous phase) into a non-polar organic solvent (liquid 2) and (ii) to facilitate/accelerate the addition reaction to a given electrophile E^+ [STA 71].

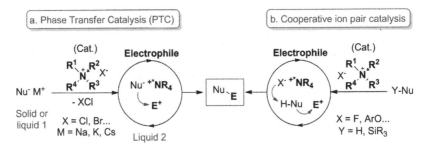

Scheme 2.1.

In 1984, in a program aiming at developing an asymmetric synthesis of new uricosuric (+)-indacrinone, the pharmaceutical company Merck described an enantioselective alkylation of indanone **1a** in the presence of 10 mol% of a chiral cinchonine-derived quaternary ammonium salt **3a** (Scheme 2.2) [DOL 84]. Through liquid/liquid biphasic mixture (toluene/50% aqueous NaOH solution), this organocatalyst, derived from natural alkaloids of *Cinchona*, furnished the corresponding methylated-compound **2a** with more than 90% *ee*. Although previous observations of non-racemic chemical transformations were reported regarding the influence of chiral ephedrinium or cinchona ammonium halides [COL 75, HIY 75, FIA 75, HEL 81], this first highly enantioselective contribution paved the way for further applications of *Cinchona*-based ammonium salts as competent asymmetry promoters for PTC.

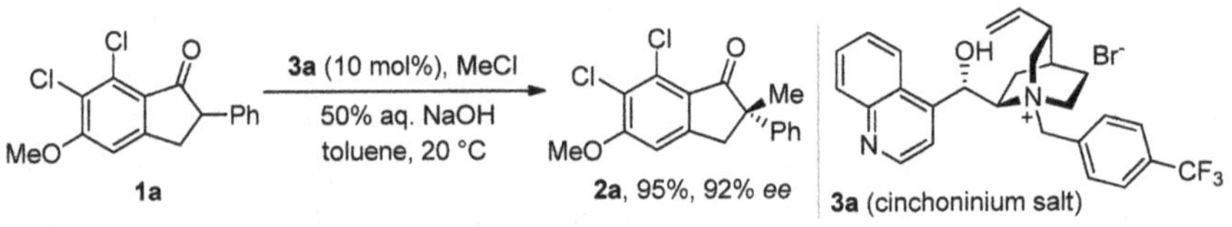

Scheme 2.2.

Five years later, the O'Donnell group of communicated on a research program concerning the alkylation reaction of imine derived from glycine *tert*-butyl ester **4a** upon PTC conditions (Scheme 2.3) [ODO 89, ODO 04]. Thereby, the corresponding (*R*)-enantioenriched benzylated-product **5a** was readily formed by means of *N*-benzyl cinchoninium catalyst **3b**. Importantly, this group demonstrated that the *pseudo*-enantiomer catalyst **6a**, synthesized from the naturally occurring cinchonidine, gave the (*S*)-benzyl glycine-ester derivative **5a** with similar *ee* but inverse enantioselectivity. This chemical transformation not only allowed for a general and straightforward access to various pharmaceutically relevant α-amino acids [MAR 07b, MAR 10b], but also became a benchmark reaction for the evaluation of new PT catalysts [HAS 07, JEW 09, SHI 13a, OOI 07b, HER 14]. Since then, many groups of research have embarked on the expension of asymmetric PTC meanwhile improving the efficiency of new chiral ammonium salts. Up until the end of 2000s, the design of more competent *Cinchona*-based ammonium salts highlighted the *N*-anthracenylmethyl derivatives from the groups of Lygo and Corey followed [LYG 04b], later on and among others, by the dimeric architectures developed by the groups of Jew and Park [JEW 09] among others [NOV 13]. In 1999, Maruoka opened a new avenue in the history of PTC by introducing highly efficient axially chiral ammonium salts [OOI 07b, SHI 13a].

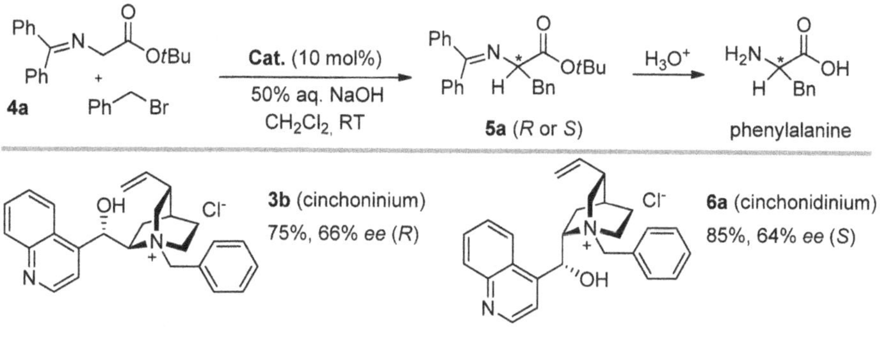

Scheme 2.3.

The substantial research surrounding PTC occulted, for a while, that chiral quaternary ammonium salts might play a synergistic action with its counterion in asymmetric catalysis. The premise of this concept, the currently so-called cooperative ion pair organocatalysis (Scheme 2.4) [OOI 04a, BRI 12, GOD 15], was revealed by Colonna, Hiemstra and

Wynberg as early as 1978 during the study of a Michael conjugated addition of nitromethane to chalcone **7a** (Scheme 2.4) [COL 78]. Making use of the *in situ* formed quininium fluoride catalyst **9a,F**, or the *N*-methylephedrinium homologue **10,F**, Wynberg and co-workers stated that *"aminium fluoride salts in which the fluoride ion serves as the base and the aminium cation functions both as a solubilizing agent for the fluoride ion in the aprotic medium and as a supplier of chirality"*. This explained the origin of the enantioselectivity observed for product **8a** and, thereby, have clearly set the stage for cooperative ion pair organocatalysis.

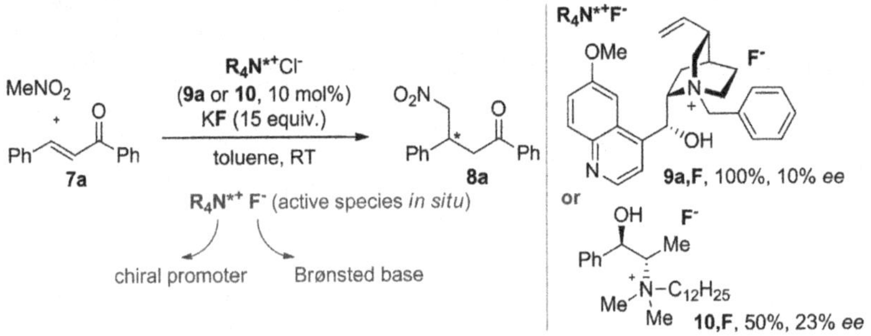

Scheme 2.4.

Since 1990, the cooperative ion pair organocatalysis has steadily grown with various modes of activation, which encompass the use of more elaborated $R_4N^{*+}X^-$ entities. Efficient dual catalysts emerged from *Cinchona*-based ammonium salts to axially chiral derivatives possessing various counter-cations (F^-, PhO^-, etc.) acting either as a Brønsted base or a Lewis base (Scheme 2.1(b)) [OOI 04a, BRI 12, GOD 15]. More recently, betaine catalysts, displaying a zwitterionic architecture, turned out to be efficient "intramolecular ion-paired organocatalysts" [NOV 13, GOD 15].

Today, chiral ammonium salts are standing at their own place in organocatalysis with regard to a great deal of reported applications. This chapter aims to provide an overview of this active field of research with, hopefully, selected but pertinent examples. It is not our goal to be exhaustive but, instead, to guide or elicit the curiosity of readers interested in this field of research. Nonetheless, the precedent reviews that have covered specific topics of these research endeavors [MAR 08b, SHI 13b, NEL 99,

OOI 07b, HAS 07, JEW 09, SHI 13a, HER 14, NOV 13, OOI 04a, BRI 12, GOD 15], together with papers dealing with industrial applications [MAR 08a, IKU 08, TAN 15], will also be cited along the way. This chapter will start with the applications of chiral ammonium salts and their properties in PTC with regard to the nature of the bond, which has been created. Then, cooperative ion pair organocatalysis will be considered.

2.2. Phase transfer catalysis

PTC usually entails user-friendly procedures and typically occurs under mild reaction conditions. Excepts for the use of quite elaborated enantiopure phase transfer catalysts in modern applications, these procedures often make use of environmentally benign and inexpensive reagents such as mineral bases, which facilitate the extrapolation to larger scale syntheses in industry [MAR 08a, IKU 08, TAN 15]. Worthy of note is the unique capability of PTC, by means of metal hydroxides (NaOH, KOH) and in specific cases, to replace the use of strong and highly moisture sensitive organic bases such as alkyl or amide-lithiums when sufficiently acidic substrates are employed. Nevertheless, mechanisms in action alongside asymmetric PTC involve various and equilibrated reaction events before the enantioselective bond formation step.

2.2.1. Phase transfer catalyst: properties and mechanism

PTC, promoted by a quaternary ammonium salt particularly, is a unique field of research within organocatalysis. This process encompasses both the formation of a chiral ammonium-nucleophile ion pair ($R_4N^+Nu^-$: organocatalytic species) and the phase transfer events of reactive species between two insoluble liquid/liquid or liquid/solid phases (Scheme 2.5(a)) [MAR 08b, SHI 13b, OOI 07b]. The reactive ionic species $R_4N^+Nu^-$ are formed via an equilibrated ion metathesis process either from the anionic reagent Na^+Nu^- or after the deprotonation of an acidic nucleophile Nu-H with a suitable mineral base (NuH + NaOH). This sequence leads to the formation of an "activated" anionic nucleophile flanked by a large and lipophilic ammonium cation much more reactive and soluble than the metal counterpart. As exemplified in Scheme 2.5(b) with a bromide anion instead of a nucleophile, the coulombic interaction energy is inversely proportional to the ionic radii that are correlated to the cation size. Generally speaking, the decrease in electrostatic energies leads to less tight ion pairs, which is correlated with the higher reactivity of the anionic nucleophile. Moreover, the lipophilic character of the ammonium moiety will facilitate both the extraction and the solubilization of the polar anionic nucleophile into the apolar organic liquid phase. This phenomenon accounts for the acceleration of the reaction rate for a given nucleophile – more than 1,000 times in certain cases – in the presence of amphiphilic quaternary ammonium salts.

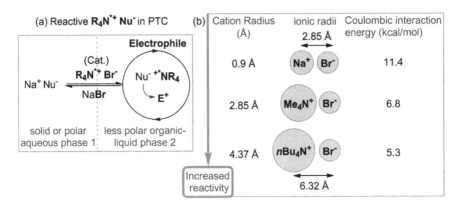

Scheme 2.5.

Nevertheless, the chiral induction that is generated by the action of chiral ammonium salt may become much more complicated to rationalize [RAB 86, DEN 11a, DEN 11b, DE 13]. First of all, one has to take into account that the non-symmetrical shape of chiral R_4N^{*+} may induce different approaches of the anion. For example (Scheme 2.6(a)), based on the analysis of the X-ray structure of a N-(anthracenylmethyl)-cinchonidinium p-nitrophenoxide, Corey and co-workers proposed a general model for Cinchona alkaloid PT catalysts whereby the counter-cation (actually the anion or the anionic nucleophile) lies at the proximity of the most accessible front-face of a virtual tetrahedron drawn from the four carbon surrounding the N^+ atom of the quinuclidinium heterocycle. Actually, the three other faces of this virtual tetrahedron are shielded by the N^+-alkyl and O-allyl moieties, and the bicyclic backside of the quinuclidine part [COR 97]. Next, beside the main electrostatic attraction originated from R_4N^+, possible secondary π–π stacking with aromatic moieties and/or hydrogen bonding with C9-alcohols create extended interactions with the nucleophile (Scheme 2.6(b)). The most intriguing feature of ammonium cations is their ability to form hydrogen bonding through the **C–H** bonds at the α-position of

N^+, which leads to a rather organized supramolecular structure with regard to the anionic nucleophile, as demonstrated by X-Ray diffraction analyses and theoretical calculation with enolates (Scheme 2.6(c)) [REE 93, REE 99, CAN 02, SHI 15]. Then, it is understood that intimate and various interactions might orchestrate the organization of the ammonium ion pair structure not only at the ground state, but also into the tertiary structure with both the nucleophile and the electrophile within the enantiodetermining transition state [DE 13, GOM 08].

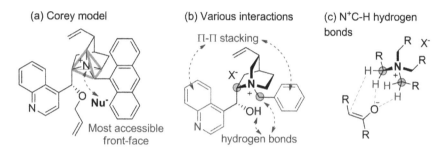

Scheme 2.6.

The phase transfer events of reactive species between the two liquid/liquid or liquid/solid phases add another catalytic dimension in the world of PTC [MAR 08b, SHI 13b, OOI 07b]. Hence, the rationalization of these complex reaction mechanisms elicited dedicated and insightful discussions [RAB 86, YAN 03, DEN 11a, DEN 11b, OOI 07b]. Generally speaking, the phase transfer events occur either through the interfacial mechanism (Makosza proposal) or the extraction mechanism (Stark proposal). As illustrated in Scheme 2.7(a), the gradual deprotonation of glycine imine **4a** (**NuH**) by an excess of NaOH leads to the ammonium-enolate ion pair **4a'**, after extraction by the PT organocatalyst ($R_4N^+Br^-$). Then, the chiral ammonium salt allows for the differentiation of the two prochiral faces of the enolate in order to secure the preferential formation of the (*R*)-benzylated product **5a** – by shielding the backward face. Note that the control of the alkene geometry, namely the *E* configuration versus the *Z* configuration, is crucial for the stereoinduction. This issue is usually controlled because of the balance between both the enolate stability and specific interactions with the ammonium cation [ODO 04, LYG 04b]. This benchmark reaction in PTC is considered as a representative example of the Makosza's interfacial mechanism whereby the two reactants (**NuH** and BnBr) are lying into the organic phase (Scheme 2.7(b)). Accordingly, the deprotonation of the nucleophile occurs at the interface between the organic and the aqueous layers containing the base to provide a small amount of sodium enolate (**Nu⁻Na⁺**). Then, an ion-metathesis reaction takes place at the interface with the amphiphilic quaternary ammonium salt **R₄N⁺Br⁻** to give **R₄N⁺Nu⁻** and allows the extraction of the nucleophilic anion into the organic phase. Subsequently, the nucleophilic substitution process takes place and the PT catalyst **R₄N⁺Br⁻** is regenerated. Likewise, a solid/liquid interfacial process occurs by means of a suspension of a solid base. This system also takes into account that small and highly hydrated anions such as OH⁻ are unlikely to be extracted into the organic layer. The optimization of these processes can make use of solid or aqueous solutions of hydroxides, carbonate or phosphate bases with various metallic cations (Na, K, Cs and Rb) covering a large array of pK_a (Scheme 2.7(c)). Saturated aqueous solutions are often used in order to maximize the empirical basicity of the solvated base together with facilitating the extraction process. Immiscible apolar solvents are usually preferred in order to favor the **R₄N⁺Nu⁻** interaction, although larger possibilities are allowed with solid/liquid processes. In this mechanism, the mass transfer may be the rate-determining step so that a high stirring rate (> 200 rpm) is required to increase the interfacial area in order to accelerate the process.

Example of Makosza interfacial mechanism

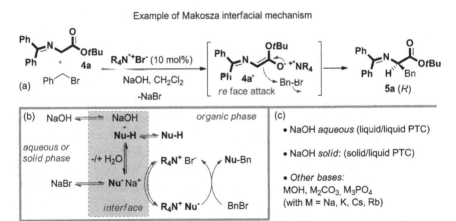

Scheme 2.7.

As depicted in Scheme 2.8(a), a PT catalyst is also able to promote the enantioselective addition reaction of an achiral nucleophile such as sodium hypochlorite (Nu^- = OCl^-) to chalcone **7a**. Contrary to the stereo-differentiation of the two prochiral faces of an enolate (Scheme 2.7(a) versus 1.8(a)), this asymmetric epoxidation reaction demonstrates the capability of the $R_4N^{*+}Nu^-$ ion pair to differentiate the enantiotopic faces of an uncharged electrophile. Furthermore, this other scenario displays a charged nucleophile lying into one phase (aqueous solution or solid phase), meanwhile the electrophile (chalcone **7** in this example) is dissolved into the second organic phase. Such processes can follow the so-called Stark extraction mechanism [STA 71], whereby the concentration of the ammonium salt $R_4N^{*+}Cl^-$ is distributed between the two phases (Scheme 2.8(b)). Then, the ion metathesis sequence takes place into the aqueous phase to furnish the $R_4N^{*+}Nu^-$ ion pair, which is extracted into the organic phase to trigger the epoxidation reaction. On the other hand, the Brändström–Montanari modification proposes a similar pathway (Scheme 2.8(c)), although the ion metathesis step occurs at the interface of the two phases. The anionic nucleophilic species is in equilibrium between the phase boundary and the aqueous layer flanked by a sodium cation, or with the organic layer having an ammonium cation. Upon the Stark extraction mechanism, the bond forming reaction between Nu^- and E^+ is often the rate determining factor; a step facilitated when the catalyst lipophilicity increases. Despite an efficient stirring is usually preferred in PTC, this is a less important issue in Stark extraction mechanism than in Makosza interfacial mechanism.

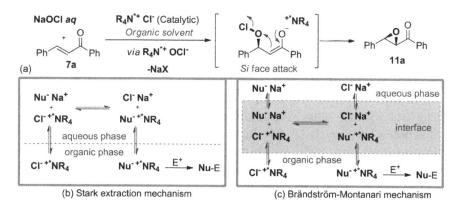

Scheme 2.8.

A balance between the hydrophilic and lipophilic characteristics of the ammonium salts as catalyst has to be found for a specific reaction based on the nitrogen charge accessibility (shape of the catalyst) and the alkyl chain sizes [RAB 86, CHI 07]. For instance, a hydrophilic ammonium salt such as triethylbenzylammonium salt is preferred when a hydroxide-initiated PTC is operating upon interfacial mechanism. One has to take into account, however, that the design of an efficient chiral architecture may complicate the full optimization of the amphiphilic catalyst, as beautifully discussed by Denmark and co-workers during the structure–activity optimization of new PT ammonium catalysts [DEN 11a, DEN 11b]. With regard to the mass transport process, we consider that the interfacial mechanism tends to occur with nucleophile having pK_a between 16 and 23, which is usually encountered for the substitution reaction of enolate along with hydroxide deprotonation events. On the other hand, acidic nucleophiles will tend to react upon the extraction mechanism. Nonetheless, due to the occurrences of preequilibrated events, one has to take into account that the distinction between the extraction and the interfacial mechanisms is not always clear-cut and both processes can also coexist.

2.2.2. Chiral catalysts: an overview

Insightful reviews have been recently written for those who want a detailed description concerning the development of chiral ammonium catalysts useful in PTC [OOI 07b, SHI 13a, NOV 13, JEW 09, MAR 08b, KAN 16]. A historical overview will be given in this chapter with representative examples of PT catalysts. Following the preliminary investigations with ephedrinium salt [COL 75, HIY 75, FIA 75], the N-benzyl ammonium salts derived from Cinchona alkaloids [HEL 76], turned out to be privileged backbones of PT catalysts because of the seminal contributions of Merck researchers and O'Donnell (Scheme 2.9) [DOL 84, ODO 04]. These salts are easily synthesized by benzylation of the quinuclidine heterocycle of the chiral-pool derived chiral amines extracted from Cinchona plant and bark [MAR 10a, YEB 11]. They are provided as four analogues divided in two pairs of pseudo-enantiomers, namely, quinine (QN) versus quinidine (QD) and cinchonidine (CD) versus cinchonine (CN), due to the fixed absolute configuration of the alkene substituted stereogenic center. Accordingly, the so-called pseudo-enantiomeric effect, namely a reverse asymmetric induction making use of quinine instead of quinidine for instance, is usually but not systematically observed. This might complicate the choice of the good structure in certain cases. Nevertheless, these reasonably cheap and readily available amines allowed for the elaboration of a myriad of ammonium catalysts for decades with outstanding applications [LYG 04a, JEW 09]. Three main points of diversity have been exploited: (i) the use of N^+-benzyl derivatives having various substitution patterns, which is the main point of structural diversity; (ii) the alcohol at C9 affords either the opportunity to create hydrogen bonding with substrates or to modulate the steric hindrance after functionalization; (iii) finally, the demethylation reaction of the methoxy at C6' of quinine and quinidine allows the formation of another OH group for hydrogen bonding or to perform further functional group manipulation. Thereby, a modulation of the topology and the electronic properties are allowed [COR 97].

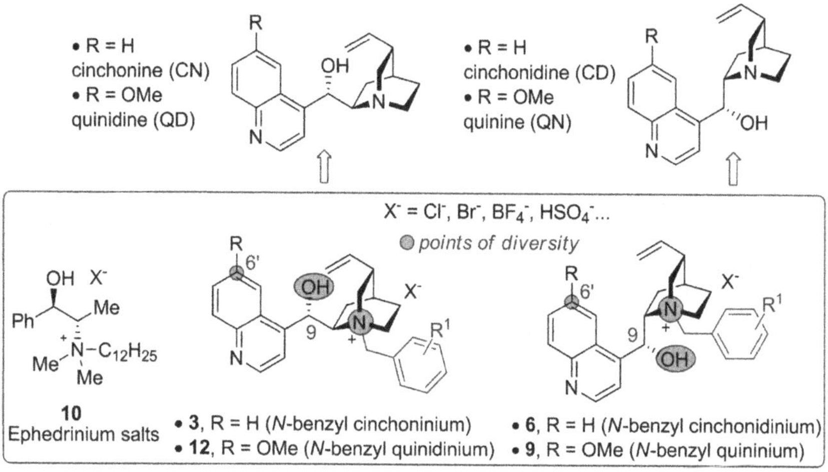

Scheme 2.9.

At first, we will make use of the benchmark benzylation reaction of glycine derivatives for comparison purposes (Scheme 2.10) [OOI 07b]. For instance, O'Donnell and colleagues succeeded in 1989 in the benzylation of glycine-imine by means of cinchonidinium **6a** with 61% *ee* (Scheme 2.10(a)) [ODO 04]. A mechanistic investigation showed that *O*-benzyl derivative **6b** formed *in situ* was actually the active catalyst [ODO 94]. In 1997, Lygo and Corey reported on an improved generation of *Cinchona* catalysts **6c–d** featuring an *N*-anthracenylmethyl moiety (Scheme 2.10(b)) [LYG 04a, COR 97]. This bigger pendant favors a better stereo-differentiation by limiting the accessibility around the chiral ammonium environment (see

Scheme 2.6(a) for details). Jew and Park triggered a systematic structure–activity relationship investigation to eventually show that a benzylic moiety displaying an aryl with a heteroatom at the 2-position led to improved enantiomeric excesses (6e–f) (Scheme 2.10(b)) [JEW 09]. The authors proposed that a molecule of water is chelated between the oxygen atom at C9 and the benzyl moiety so that the catalyst is rigidified in a more efficient conformation, as depicted with structures 6e–f. The same group of researchers proposed another design based on the connection of two or three cinchonidinine units thanks to a specific benzylic tether (Scheme 2.10(c)). Among several analogues 6h–k, the 1,3-disubstituted 6i or 1,3,5-trisubstituted derivatives 6k gave the best results [JEW 09]. In this vein, Tan recently reported on a bis-ammonium PT catalyst derived from imidazolium salts for the asymmetric oxidation reaction of alkenes [WAN 15b]. In 1999, Maruoka and colleagues opened a new route in PTC by designing axially chiral PT-catalysts such as 13 having N-spiro ammonium binaphtyl backbones (Scheme 2.10(d)) [OOI 99, OOI 07b, SHI 13a]. This C_2-symmetrical architecture requires more synthetic effort to be elaborated, but could be used at catalyst loading as low as 0.01 mol% and offer many salient features: (i) an increased stability due in part to the absence of hydrogen atom at the β-position of the nitrogen atom, which prevents the Hoffman elimination in basic conditions; (ii) the possibility to modulate the steric, electronic and lipophilic properties through the introduction of various aryl moieties at 3 and 3′ positions. Actually, the nature of aryl pendants at 3 and 3′ positions are at the origin of the efficient stereo-differentiation by creating a chiral pocket. For instance, the glycine alkylation reaction proceeded with almost perfect enantiomeric excess as long as trifluorophenyl substituents (13d versus 13a–c) are connected to the binaphtyl backbone [OOI 99, OOI 07b, SHI 13a]. Maruoka proposed a second and more accessible generation of catalysts, which only required a dibutyl pendants in order to furnish high enantiomeric excesses (14a, Scheme 2.10(e)) [KIT 08]. Interestingly, the same group demonstrated that spiranic ammonium catalyst 15a displaying an axially flexible biaryl unit, which is a readily available and less costly moiety, could be used upon the influence of a well-defined binaphtyl moiety (Scheme 2.10(e)) [OOI 02c, OOI 06c]. Then, a mixture of homochiral (S,S)-15a and heterochiral (R,S)-15a of the so-called $tropos$-catalyst exists in solution but the homochiral (S,S)-15a turned out to be a much-more reactive and selective species in catalysis [IUL 10]. As far as Lygo is concerned, a new kind of $tropos$-catalyst 16 was designed based on a central-axial chirality communication, which could be optimized by a convenient *one-pot* construction [LYG 03, LYG 04a, LYG 09a].

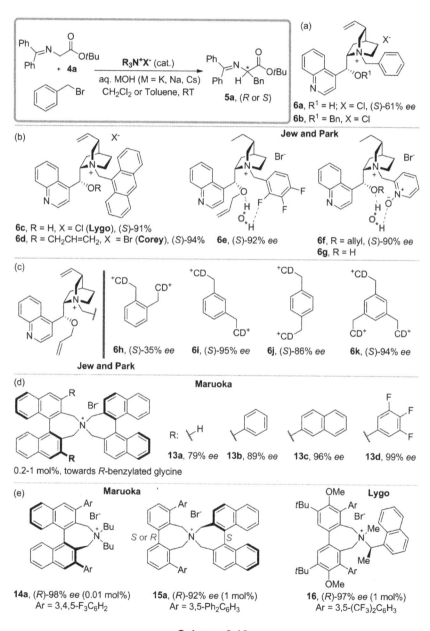

Scheme 2.10.

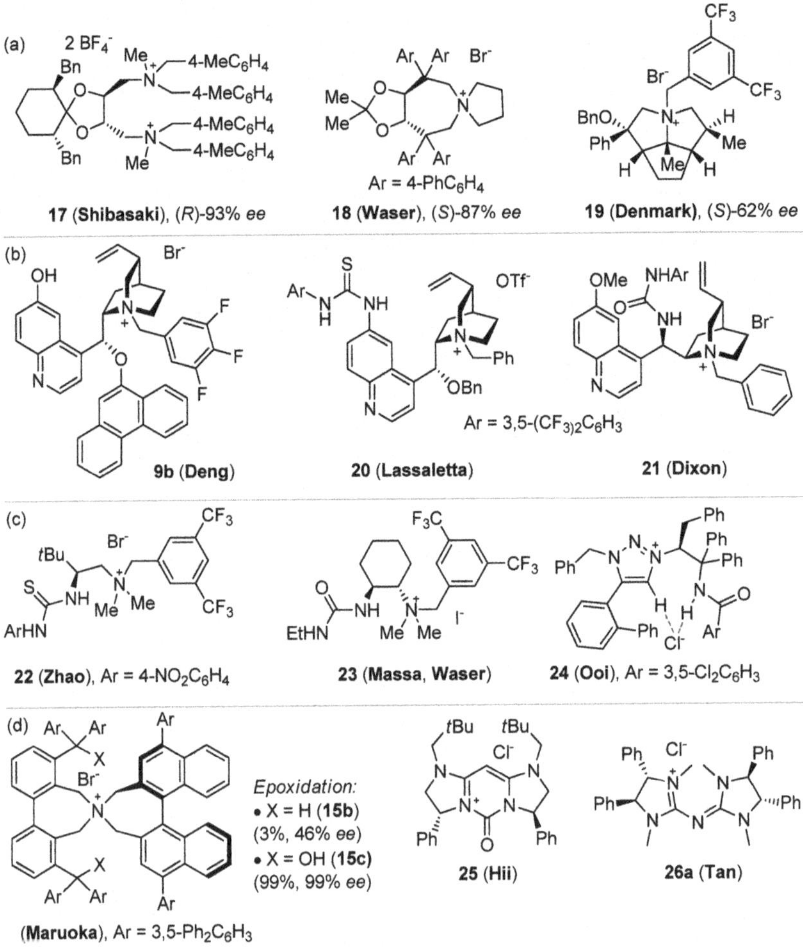

Scheme 2.11.

Recently, a new family of chiral quaternary ammonium salts emerged in the literature (Scheme 2.11) [OOI 07b, SHI 13a, NOV 13]. Although they are promising and feature high levels of performance in certain cases, they have not yet been evaluated on a large array of reactions in PTC, on the contrary to native *Cinchona*-derived catalysts or spiro-ammonium salts developed by the group of Maruoka (Scheme 2.10). Making use of tartaric acid, as readily available starting material, Shibasaki and colleagues proposed a novel generation of bis-ammonium salt **17** highlighting the concept of two-center PT catalyst for alkylation reaction of glycine derivatives

(Scheme 2.11(a)) [SHI 02, OHS 04]. In this study, the authors also pointed out the significant influence of the counterion on the efficiency of the process [OHS 04]. Waser reported on the synthesis of N-spiro quaternary ammonium salt **18** obtained from tartaric-derived TADDOL backbone [WAS 12, GUR 13]. Denmark and co-workers published an impressive and insightful structure–activity relationship investigation by means of a modern chemoinformatic analysis of novel tricyclic ammonium salts such as **19** [DEN 11a, DEN 11b]. These types of compounds **19** were obtained through a diversity-oriented synthesis approach because of tandem cycloaddition reactions of nitroalkene allowing the identification of catalyst **19**. Recently, novel bifunctional ammonium salts, namely having an another functional group able to form extended stabilization interactions beside the ammonium moiety, have emerged as promising PT catalysts on various types of reactions [NOV 13]. The design of these PT catalysts has explored the possibilities to make use of hydrogen bond donor functionalities different than the original alcohol at C9. Berkessel, for an epoxidation reaction in 2007, and Deng, for an asymmetric Darzens reaction in 2011, demonstrated the advantage of having a hydrogen bond donor element on the quinoline ring [BER 07, LIU 11]. This was exemplified with quinine derivative **9b**; in which the methoxy functional group was demethylated (Scheme 2.11(b)). Subsequent to the work of Lassaletta and Fernández with thiourea-quinine catalyst **20** [BER 10], the groups of Dixon [JOH 12], Smith [LI 13] and Lin and Duan [WAN 14, WAN 15a] showed that the presence of urea and squaramide functional groups at C9 afforded efficient bifunctional catalysts such as quinidine derivative **21** for intra- and intermolecular alkylation or fluorination reactions. Zhao *et al.* (**22**, Scheme 2.11(c)) [WAN 13a, WAN 13b], and more recently Jiang and colleagues [DUA 15], described bifunctional thiourea-ammonium derived from readily available α-amino acids as efficient PT catalyst for Mannich-type reactions. In the same vein, Waser and Massa, respectively, have shown that *trans*-1,2-cyclohexanediamine served as a useful backbone for the design of thiourea-ammonium PT catalyst such as **23** [NOV 14, PER 14, TIF 15]. Recently, Ooi and co-workers demonstrated by ^{1}H NMR and X-ray diffraction the capability of a novel triazolium salt **24** to recognize chloride anion through hydrogen bond interactions [OHM 11]. Based on this principle, the ammonium salt **24** turned out to be an efficient catalyst for the enantioselective alkylation of oxindoles. Maruoka pursued the investigation of *tropos-spiro* ammonium salts such as compound **15b** and took advantage of the readily available biphenyl moiety to introduce alcohol functionalities (Scheme 2.11(d)). The importance of these alcohol functional groups for getting high enantiomeric excess in chalcone epoxidation reaction was demonstrated for instance [OOI 04d, NOV 13]. Hii highlighted the capability of original 2-oxopyrimidinium salt **25** to perform as an efficient PT catalyst for the 1,4-addition of glycine imine to enones. Tan and colleague have recently introduced recently a novel family of PT catalyst **26a** displaying a pentanidium backbone for low loading phase transfer catalytic reactions involving C–C, C–O and C–S bond formation [MA 11, YAN 12, ZON 14]. With pentanidium derivatives **26a**, 2-oxopyrimidinium salt **25** and triazolium **24**, Tan, Hii and Ooi highlighted that, besides sp^3 quaternary ammonium salts, planar sp^2-hybridized ammonium salts can also be exploited for the design of novel efficient PT catalysts. In summary, the beginning of asymmetric phase transfer catalytic processes was dominated by quaternary ammonium *Cinchona*-derived catalysts up to early 2000s [JEW 09]. Maruoka and co-workers have written a new chapter with efficient *N*-spiro-ammonium derivatives **13–15** having an axial chirality [OOI 07b]. Among different interesting catalyst design [OOI 07b, SHI 13a], bifunctional ammonium salts [NOV 13], flanked by a functional group capable of donating hydrogen bonds, or sp^2-hybridized ammonium salts have emerged in new asymmetric synthetic application [KAN 16].

2.2.3. C–C bond formation

The enantioselective construction of C–C bonds is by far the oldest and most productive research endeavor in PTC. It would be impossible to exhaustively cover the great deal of achievements in this chapter. Recent insightful reviews have already beautifully described this field of research [OOI 07b, JEW 09, SHI 13a, HER 14, MAR 08b, NOV 13, KAN 16]. Accordingly, we prefer herein to provide an overview through chosen examples and, thereby, to highlight the capability of advanced chiral PT catalysts to promote enantioselective C–C bond formation reaction and concomitantly bringing up some functional group tolerance issues. As already described in Chapter 1 (see Schemes 2.10 and 2.11), the benzylation or allylation of glycine imines **4a** served as a benchmark reaction for the development of new and efficient PT catalysts. For instance, the usefulness of this synthetic method toward the formation of various α-amino acid derivatives **5a** has been established with almost perfect *ee*s by means of the Maruoka axially chiral *spiro*-ammonium salt **13d** (with 50% aq. KOH) not only with benzyl and allyl bromide electrophiles (**E-X**) but also with saturated alkyl iodide with the use of saturated aqueous CsOH at lower temperature (Scheme 2.12(a)) [OOI 99, OOI 07b, SHI 13a]. The same group also highlighted the kinetic resolution of secondary alkyl bromide **27a** with 95% *ee* and 95/5 d.r. accelerated in the presence of a crown ether additive (Scheme 2.12(b)) [OOI 07a]. In this case, catalyst **13e** displaying sterically hindered aryl substituents at 3,3'-position was required for getting high *ees* together with the help of a crown ether, which enhances the reactivity of the overall system. It is believed that the crown ether acts as an achiral PT co-catalyst extracting potassium hydroxide into the organic phase in order to facilitate the deprotonation of glycine imine.

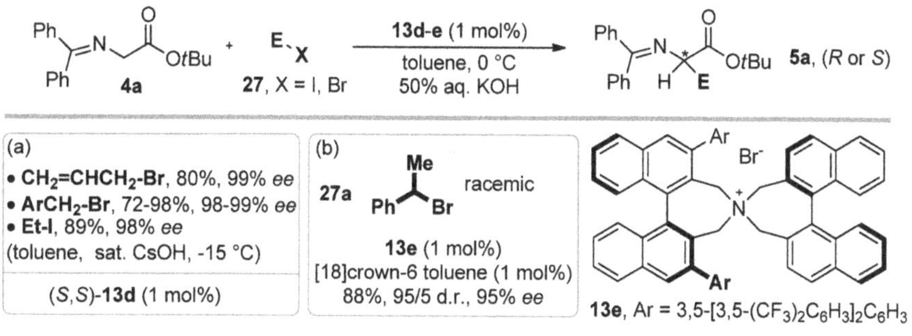

Scheme 2.12.

Making use of the Corey–Lygo PT-catalyst **6d**, the introduction of the iodo-epoxide **27b** was achieved to construct the corresponding alkylated glycine **5a** as a single diastereoisomer (Scheme 2.13(a)) [BOE 02]. The authors made use of this precursor in order to synthesize an amino caprolactam **28** as an advanced intermediate toward the elaboration of naturally occurring bengamides. The enantioselective alkylation of glycine imine **4a** by propargylic bromide **27c** furnished the corresponding product with 94% *ee* by means of the Jew–Park PT catalyst **6e** (Scheme 2.13(b)) [CAS 03]. The alkyne pendant was transformed into the indol derivative **29** in few steps, which is the central tryptophan residue of the bicyclic octapeptide Celogentin C. During the elaboration of a key intermediate of a drug candidate, GlaxoSmithKline tackled the kilogram scale synthesis of an unnatural amino acid, namely the 4-fluoro-β-(4-fluorophenyl)-L-phenylalanine **30** (Scheme 2.13(c)) [PAT 07]. It was demonstrated that the alkylation of glycine Schiff base **4a** by electrophilic bromide **27d** proceeded with 60% *ee* (99% *ee* after one recrystallization with 56% yield) in the presence Corey–Lygo PT catalyst **6d** allowing to provide **30** after acid hydrolysis of the imine moiety.

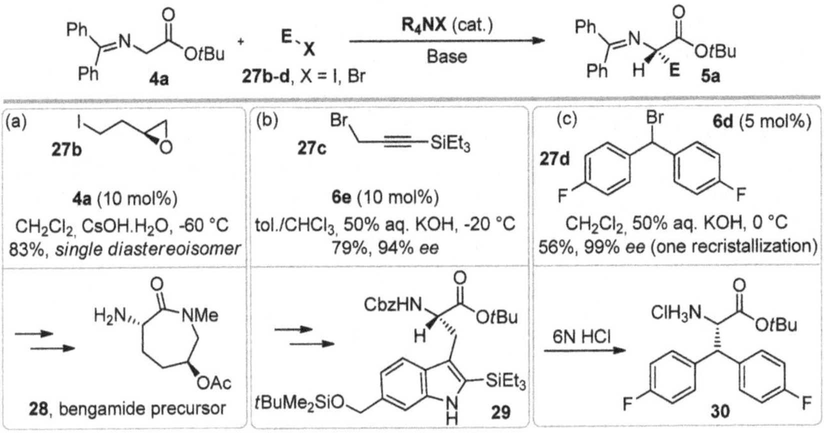

Scheme 2.13.

The conjugate addition reaction onto electron-poor alkenes is not only a key reaction in asymmetric organic synthesis but also served as a useful platform for the development of PT catalytic conditions. One has to take into account in such a process that the deprotonated glycinate **4a** adds first to the Michael acceptor **31** to furnish the anionic intermediate **32** (Scheme 2.14). Then, the anion **32** must be rapidly protonated in order to prevent any retro-Michael reaction that would lead to an overall thermodynamically controlled process and a racemic product **5a**. In practice, the anion **32** traps a proton from a protic solvent (usually water), the protonated base (use in catalytic quantity in this case) or the starting material **4a** with a suited pK_a value ($<pK_a$ of **32**). In the last case, a phase transfer initiated process occurs and the catalytic cycle does not require any more base afterwards. For instance, the enantioselective conjugated addition of glycine Schiff base **4a** has been achieved to acrynonitrile **31a** with Corey–Lygo catalyst **6d** to open an access to naturally occurring (*S*)-ornithine (Scheme 2.13(a)) [ZHA 00]. The addition to acrylate **31b**, vinylsulfone **31c** and other substrates was allowed by means of non-ionic base BEMP **33** in the presence of Jew–Park catalyst **6e** (Scheme 2.14(b)) [ODO 01]. The group of O'Donnell extrapolated these homogenous conditions to the alkylation of glycine **4a** on solid support [ODO 01]. In the same vein, Shibasaki and co-workers performed the 1,4-addition reaction to acrylate in biphasic solid/liquid conditions with the

tartaric acid derived bis-ammonium PT catalyst **17** [SHI 02]. By means of only 1 mol% of his *tropos*-catalyst **16** and 0.5 equiv. of base, Lygo succeeded in performing the akin 1,4-addition reactions to enone **31d-e** with high *ees* (Scheme 2.14(c)) [LYG 05].

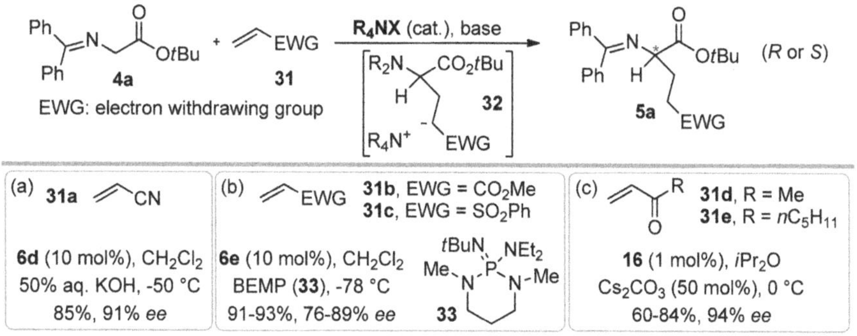

Scheme 2.14.

Numerous applications of the asymmetric PT catalytic conjugated addition have been published and successfully used in target oriented synthesis [OOI 07b, MAR 08b, SHI 13a]. To cite a few, the 1,4-addition reaction of glycine imine **4a** has been extrapolated to β-substituted enones or enals (Schemes 2.15(a) and (b)) [HUA 11, MA 11]. Various nucleophiles were also successfully employed either to create a tertiary stereogenic center from an α-hydroxy ketone (Scheme 2.15(c)) [AND 08] or to generate a quaternary stereogenic center with pro-nucleophiles able to stabilize a transient anion after deprotonation (Schemes 1.15(d–h)) [WAN 07a, FUR 08, HUA 10, HAS 10, SHI 11]. An interesting example also proved the ability of PTC to perform the 1,4-addition reaction of a cyano malonate to an acetylenic ester, namely having a C–C triple bond instead of an activated double bond (Scheme 2.15(d)) [WAN 07a].

As depicted in Scheme 2.15, PTC is a powerful approach to construct quaternary stereogenic centers though the enantioselective alkylation of a tri-substituted stabilized anion flanked by a chiral ammonium cation. This strategy was developed as early as 1992 by O'Donnell for the elaboration of nonproteinogenic α,α-disubstituted amino acids useful as building blocks in medicinal chemistry (Scheme 2.16) [ODO 92]. The authors modified the steric and electronic properties of the Schiff base **34** by using a glycine derived from a *para*-chlorobenzaldehyde that facilitated the second alkylation event. Furthermore, they pointed out the superiority of the mixed K$_2$CO$_3$/KOH solid base in the presence of N-benzyl cinchoninium catalyst **3b** to give the corresponding quaternary amino acid precursor **35** with promising 44% *ee*. In line with this pioneer investigation, the selectivity of this synthetic sequence was subsequently improved by several groups by means of the Corey–Lygo catalyst **6d** [LYG 99a], Maruoka axially chiral catalyst **13d** [OOI 00], Jew–Park ammonium salt **6f** [JEW 03], and Shibasaki diammonium PT catalyst **17** [OHS 04].

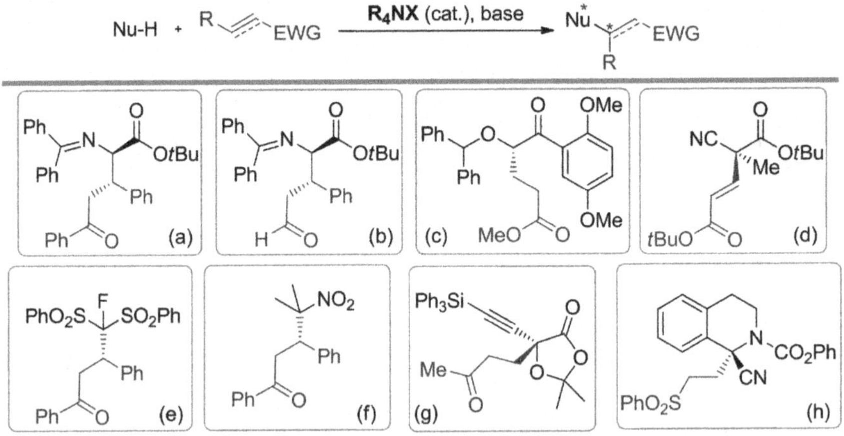

Scheme 2.15.

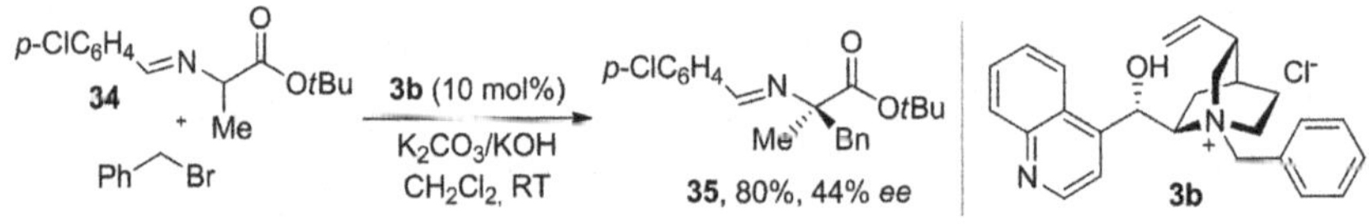

Scheme 2.16.

Likewise, the synthesis of various compounds was realized along with the control of the absolute configuration of an all carbon quaternary center (Scheme 2.17). The alkylation of cyclic derivatives allowed the construction of α-amino acid derivatives flanked by an ether, sulfide or amine pendants as precursors of homologues of serine (Scheme 2.17(a)) [JEW 04, LEE 05], cysteine and diamino-propionic acid, respectively (Scheme 2.17(b)) [KIM 06, PAR 09], as well as the homologated counterparts (Scheme 2.17(c))

[KIM 09]. The large scope of this synthetic approach was further highlighted by applying it to the functionalization of α-hydroxyamides (Scheme 2.17(d)) and divers keto ester derivatives (Schemes 1.17(e–g)) [OOI 06a, OOI 06b, NIB 09, PAR 12].

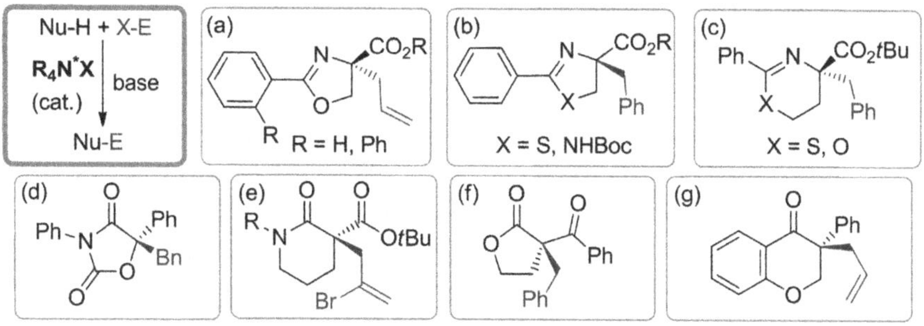

Scheme 2.17.

Jørgensen and colleagues reported on the interesting alkynylation and vinylation reactions of keto ester **1a**, showing the ability to perform a functionalization by sp^2 and sp alkyl pendants (Scheme 2.18) [POU 06, POU 07]. The strategy was based on the domino-conjugated addition–elimination reaction taking place to β-halogeno-alkynes **36** or -alkenes **31f** electrophiles. It was demonstrated that a Corey–Lygo type PT catalyst **3c** having an adamantoyl moiety on the C9 alcohol afforded the corresponding products **37–38** with *ees* ranging from 75 to 98% in many instances except for the alkynylation of a seven-membered ring giving 44% *ee*. Importantly, in case of vinylation process, the Z-alkenes **38** were mainly obtained.

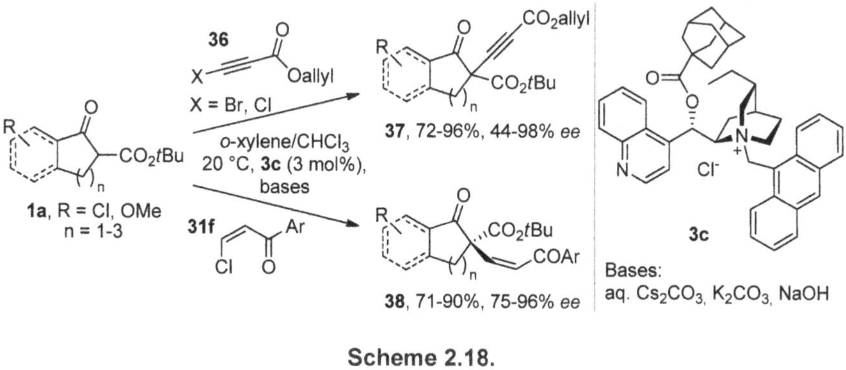

Scheme 2.18.

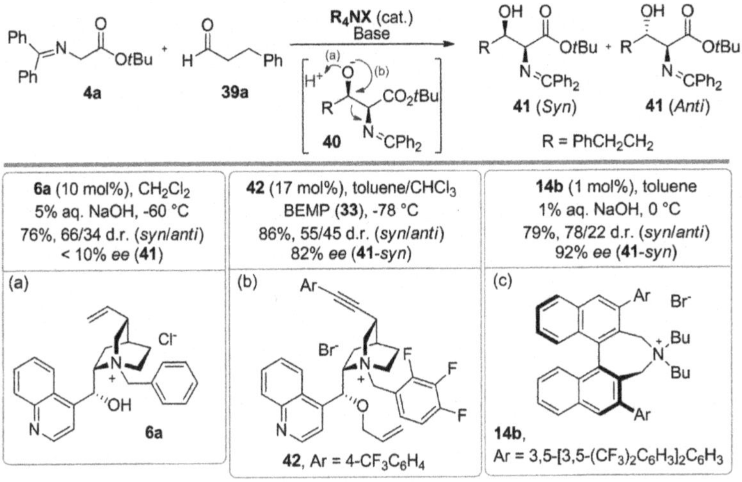

Scheme 2.19.

The PT catalytic aldolization reaction is a challenging process that met recent success, especially with glycine Schiff base **4a** (Scheme 2.19). As with the conjugated addition reaction (*vide supra*), the PT catalytic aldolization reaction leads to an alcoolate intermediate **40**, which needs to be rapidly protonated (path a) in order to prevent a retro-aldolization process to take place (path b). In the latter case, thermodynamically controlled conditions would provide a racemic alcohol **41** within moderate yields and selectivities (depending on the product stability). In 1991, Gasparski and Miller reported some pioneering investigations demonstrating that an aldolization process with glycine imine **4a** may happen and provide a useful synthesis of α-hydroxy-β-amino ester derivatives **41**. However, starting from hydrocinnamaldehyde **39a** *N*-benzyl cinchonidinium chloride **6a** and diluted aqueous solution of NaOH, the corresponding products **41** were obtained in moderate d.r. (66/34) and poor *ee* (<10%) (Scheme 2.19(a)) [GAS 91]. In 2007, Castle and colleagues evaluated a new generation of Jew and Park type PT catalyst **42** possessing an alkyne moiety (Schemes 1.19(b) versus 1.10) [MA 07]. By means of the phosphazene base BEMP **33** and dihydrocinnamaldehyde **39a** in homogeneous conditions, they succeeded in forming the *syn* aldol adduct **41** with 82% *ee* albeit with poor diastereoisomeric ratio. In 2002, Maruoka's group tackled the transformation of glycine imine **4a** into the aldol adducts **41** upon the influence of *spiro*-ammonium salt **13e** having bis-binaphtyl backbone (see Scheme 2.12 for catalyst **13e**), and obtained excellent enantioselectivies (98% *ee*) for the major *anti*-adduct **41** [OOI 02b, OOI 04c]. Interestingly, more recent investigations (Scheme 2.19(c)), upon the influence of the new generation of simplified *spiro*-ammonium salt **14b**, allowed to obtained majorly the *syn*-diastereoisomer **41** with 92% *ee* [KIT 08]. Three points of importance have been unveiled: (i) the use of a diluted aqueous solution of NaOH allows to minimize the retro-aldolization reaction, (ii) mechanistic insights revealed that the retro-aldolization event is also stereospecific and (iii) the ability to form either the major *syn* or *anti*-diastereoisomer with respect to the catalyst used (**13e** or **14b**). The asymmetric PT-aldolization process was also developed with α-diazaacetate derivatives [ARA 04].

In the same vein, the imine electrophiles **43** were used instead of aldehydes and allowed the achievement of several Mannich-type processes (Scheme 2.20). The *spiro*-axially chiral ammonium salt **6d** turned out to be a highly potent PT-catalyst to promote the enantioselective addition reaction of glycine imine **4a** to rather electrophilic α-imino esters **43a** with a *N-para*-methoxyphenyl functional group (PMP) (Scheme 2.20(a)) [OOI 04b]. By these means, the major *syn*-α,β-amino esters **44a** were obtained with high *ees* and allowed the synthesis of streptolidine lactam, an advance precursor of an antibiotic. Shortly after, Shibasaki and colleague reported on a more general Mannich-type reactions based on the use of *N*-Boc imines **43b** (Scheme 2.20(b)) [OKA 05, SHI 07]. Making use of the tartaric-derived *bis*-ammonium PT catalyst **17**, the authors obtained the major *syn*-adduct **44b** with excellent d.r. and *ees* with aromatic or aliphatic imines. Bernardi and colleagues extended this Mannich-type reaction to phosphonoglycine Schiff bases **4b** with a Jew–Park-modified catalyst **6g** (Scheme 2.20(c)) [MOM 09]. The authors generated *in situ* the required unstable *N*-Boc-imines **43b** from the bench-stable α-amido sulfones **45** under the PT biphasic basic conditions. Accordingly, the corresponding α,β-diaminophosphonic acid derivatives **44c** were synthesized as a single *syn*-diastereoisomer with *ees* ranging from 72% to 92%. Bernardi, Herrera, Ricci and Palomo took advantage of this convenient strategy to generate *N*-carbamoyl imines **43b** and successfully developed several Mannich-like reactions with nitromethane (aza-Henry reaction), malonate, sulfonyl and keto-ester nucleophiles [FIN 05, MAR 07a, GOM 08, CAS 09]. Importantly, Palomo reported an insightful theoretical investigation of PT-catalyzed addition reaction of nitromethane with *N*-Boc imines showing the importance of both the OH at C9 and the C–H bonds at the α-position of the positively charged nitrogen atom of *Cinchona*-derived ammonium catalysts in order to create stabilizing hydrogen bonding into the transition state (see Scheme 2.6 for discussion) [GOM 08]. This Mannich reaction was also a platform for the development of new PT-catalyst such as the bifunctional Dixon's catalyst **21** or the triazolium salt **24** developed by Ooi [OHM 12, JOH 12, WEI 12a]. This field of research has been very active in recent years, with a particular aim of continuously extending the scope of this useful transformation [OOI 07b, MAR 08b, SHI 13a]. Of note, the related PT-catalyzed trifluomethylation of imine electrophiles has recently been reported [KAW 09].

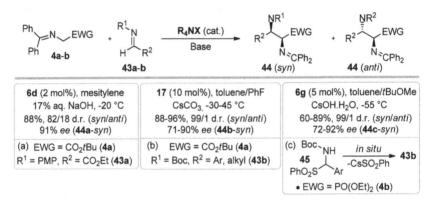

6d (2 mol%), mesitylene	17 (10 mol%), toluene/PhF	6g (5 mol%), toluene/tBuOMe
17% aq. NaOH, -20 °C	CsCO₃, -30-45 °C	CsOH.H₂O, -55 °C
88%, 82/18 d.r. (syn/anti)	88-96%, 99/1 d.r. (syn/anti)	60-89%, 99/1 d.r. (syn/anti)
91% ee (**4a**-syn)	71-90% ee (**44b**-syn)	72-92% ee (**44c**-syn)

(a) EWG = CO₂tBu (**4a**)   (b) EWG = CO₂tBu (**4a**)

R¹ = PMP, R² = CO₂Et (**43a**)   R¹ = Boc, R² = Ar, alkyl (**43b**)

(c) Boc‚NH   in situ
    45 ⟶ **43b**
    PhO₂SAr   -CsSO₂Ph

• EWG = PO(OEt)₂ (**4b**)

Scheme 2.20.

PTC has been recognized as a powerful method for the promotion of asymmetric cyclization or annulation processes wherein the Darzens and the cyclopropanation reactions standout. The recent and exhaustive review by Waser gives an good overview of the contribution of PT catalysis cyclization into this field of research [HER 14]. For example, Fini, Bernardi and colleagues described an original enantioselective formal [3+2] cycloaddition in which unstable nitrones **49** were *in situ* generated from the corresponding α-sulfone amides **46** (Scheme 2.21) [GIO 09]. The domino sequence begins with the addition reaction of the deprotonated allylic derivative **50** (from **47**) to the activated nitrone **49** and is followed by an intramolecular oxa-Michael-reprotonation sequence occurred to afford the corresponding isoxazoline **48** with very good selectivities (up to 99% *ee*). The PT catalyst **9c** was designed with a bulky group on alcohol of quinine backbone and an *ortho*-substituted benzyl moiety for reaching highest *ees*. Worthy of note, this approach smoothly worked with α-sulfone amides **46** derived from aliphatic aldehyde but the aromatic homologues remain hardly accessible.

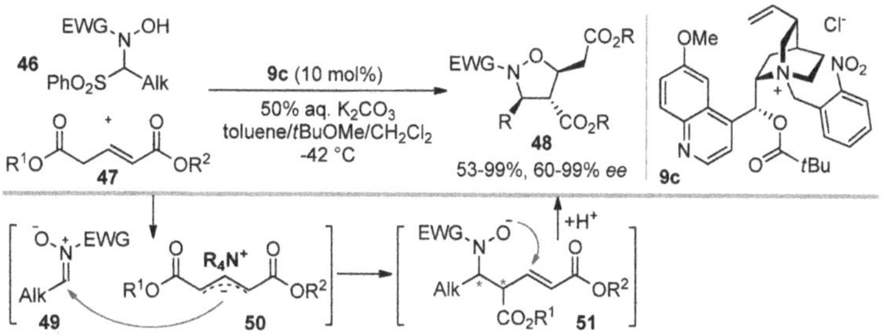

Scheme 2.21.

Smith and co-workers initiated a research program aiming at achieving organocatalyzed enantioselective electrocyclization reactions through cation-controlled processes [MAC 09, LI 13, JOH 15]. In its most recent version, the authors demonstrated that *N*-benzyl cinchonidinium ammonium salt **6a** was able to promote a base-catalyzed cyclization of imines **52** into indolines **53** with high level of diastereo- and enantio-control (Scheme 2.22(a)) [SHA 15]. A complete kinetic and theoretical mechanistic investigation showed that the pericyclic reaction may proceed from the anion **54** either by a suprafacial electrocyclic mechanism involving a 2-aza-pentadienyl anion or via a 5-*endo-trig* Mannich reaction (Scheme 2.22(b)), the former being most likely. Furthermore, it was demonstrated that a Makosza interfacial process takes place whereby the first catalytic species turned out to be the betaine **6a'** resulting from the deprotonation of the precatalyst **6a** by CsOH (Scheme 2.22(c)). Then, the zwitterionic catalyst **6a'** initiates the catalytic cycle with the deprotonation of the imine starting material **52**. Accordingly, the authors proposed the term phase transfer initiated process.

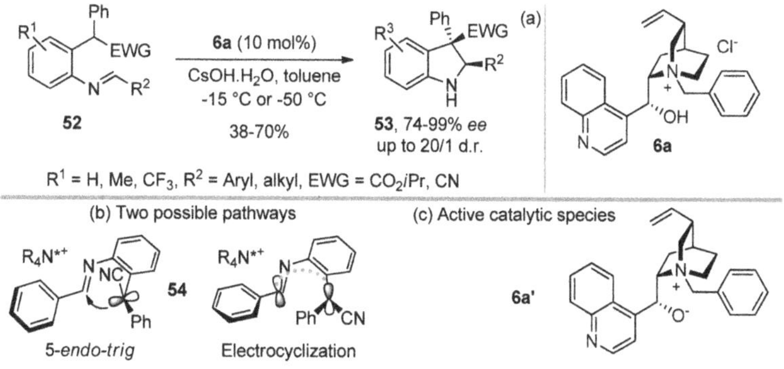

Scheme 2.22.

2.2.4. C–O bond formation

Enantioselective epoxidations of electron-deficient alkenes belong to the earliest examples of asymmetric catalysis based on the use of chiral PTCs. Their reaction mechanisms, which usually follow the Stark extraction proposition, were discussed in section 1.2.1 (Scheme 2.23). In the late 1970s, Wynberg *et al.* reported the first catalytic asymmetric epoxidation processes of chalcone **7a** and quinone **7b** under PTC by means of a catalytic amount of *N*-benzyl quininium chloride **9a** and hydrogen peroxide or sodium hypochlorite as oxidizing agents [HEL 76, HUM 78, PLU 80]. The corresponding epoxides **11a** and **11b** were obtained with modest enantiomeric excesses not exceeding 55% (Schemes 2.23(a) and (b)). In 1998, Taylor *et al.* investigated epoxidation of the readily available dienone **11b** using Wynberg's procedure as a key step in the asymmetric synthesis of (–)-Manumycin (Scheme 2.23(c)) [ALC 98, MAC 98]. Appealing levels of enantioselectivity up to 89% could be obtained using *N*-benzyl cinchonidinium chloride **6a** and *N*-benzyl quininium chloride **9a**, however, with modest yields (15–32%). Absolute configuration of the resulting (–)-epoxyquinone **11b** was assigned as *2S,3R*, although the (+)-enantiomer is needed for the synthesis of the natural (–)-Manumycin. Attempts to use the pseudoenantiomeric *N*-benzylcinchoninium chloride **3b** afforded the desired (+)-epoxyquinone **11b**, albeit in only 10% *ee*. This example points out one major limitation of using *Cinchona* alkaloid derivatives in asymmetric catalysis, namely, the difficulty to predict whether or not pseudo-enantiomers of *Cinchona* alkaloids will exhibit efficient opposite enantiocontrol.

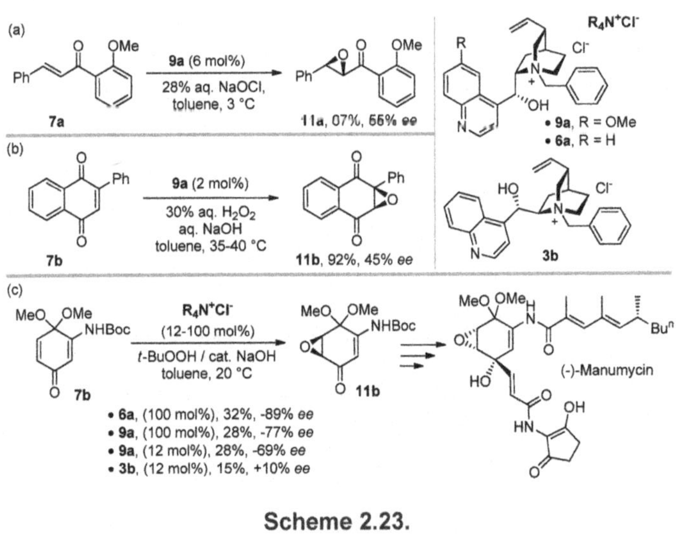

Scheme 2.23.

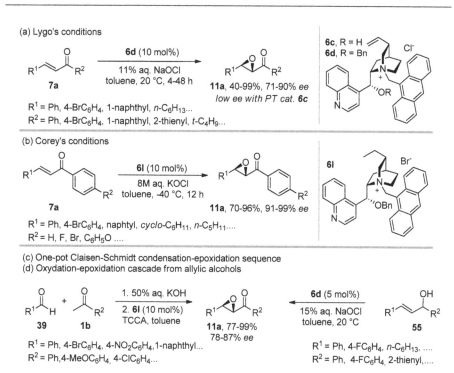

Scheme 2.24. *Enantioselective epoxidation mediated by Lygo's and Corey's PTCs*

From 1998, major strides in the phase-transfer catalyzed epoxidation of α,β-unsaturated ketones were accomplished by Lygo *et al.* and Corey *et al.* who made use of quaternary ammonium salts **6d** and **6l** bearing an *N*-anthracenylmethyl appendage (Scheme 2.24(a)) [LYG 98, LYG 99b, LYG 02, COR 99]. Generally speaking, these two catalysts afforded good yields (40–99%) and high *ees* (71–99%) with a wide range of enones **7a**. In their first papers, Lygo *et al.* highlighted the crucial role of the *O*-benzyl moiety at C-9 to reach high stereoselectivity (Scheme 2.24(a)) [LYG 98, LYG 99b]. Indeed, although PT catalyst **6c** (R = H) displayed good performances during the asymmetric alkylation of glycine-imines (see Scheme 2.10), only poor enantioselectivities were obtained in epoxidation reactions of chalcone **7a**, likely due to deleterious interactions between the oxidant and the hydroxy group of the PT catalyst **6c**. This hypothesis was substantiated by the fact that inversion of enantioselectivity is observed on switching the oxidant from sodium hypochlorite to hydrogen peroxide. One of the main difference between Lygo's and Corey's works is the oxidizing source. Although Lygo made use of sodium hypochlorite, Corey showed that potassium hypochlorite is more reactive under PT conditions, thereby allowing the reaction to occur at lower temperature (–40°C) with significantly improved enantioselectivities (Schemes 2.24(a) and (b)). In 2007, Liang *et al.* reported an elegant one-pot Claisen–Schmidt condensation-epoxidation sequence in the presence of Corey's PT catalyst **6l** and trichloroisocyanuric acid as oxidant furnishing of large variety of highly enantioenriched α,β-epoxy ketones **11a** (75–96% *ee*) from readily available aldehydes **39** and ketones **1b** (Scheme 24(c)) [WAN 07b]. Lastly, Lygo *et al.* reported on an efficient oxidation-epoxidation cascade in which allylic alcohols **55** were converted directly into α,β-epoxy ketones **7a** (78–87% *ee*) under asymmetric PTC conditions (Scheme 2.24(d)) [LYG 02]. These last two examples additionally disclose the high potential of PTC in the development of asymmetric organocascade reactions.

Another class of *Cinchona*-PTCs containing *N*-polyfluorinated- and *N*-di-trifluoromethylated benzyl moieties (**3d**, **12a–b**) also gave excellent results during the epoxidation of chalcones **7a** (Scheme 2.25). In 2010, Park *et al.* reported on the use of PTC **3d**, bearing a 2,3,4-trifluorobenzyl group, in the epoxidation of various chalcones **7a** with high level of enantioselectivities (91–95% *ee*) and good yields (71–95%) giving rise to the optically enriched (αR, βS)-epoxy ketones **11a** (Scheme 2.25(a)) [YOO 10]. In 2013, Chen reported the enantioselective epoxidation of acyclic β-trifluoromethyl-β,β-disubstituted enones **7c** with a pentafluorinated quinidine-derived PTC **12a** leading to the formation of a large panel of epoxides **11c** having a quaternary trifluoromethylated carbon center in excellent *ee* (95–99%) and high diastereoselectivities (d.r. > 20:1) at low catalyst loading (Scheme 2.25(b)) [WU 13]. In the same year, Shibata *et al.* developed an enantioselective catalytic aerobic epoxidation of β-trifluoromethyl-β,β-disubstituted enones **7c** (Scheme 2.25(c)) [KAW 13]. The required hydrogen peroxide is generated *in situ* by reacting molecular oxygen with methylhydrazine in the presence of a base. When using PT catalyst **12b**, flanked by a 3,5-bistrifluomethylbenzyl group at the quinuclidine nitrogen, excellent *ee*s were obtained (96–99%) providing an efficient access to highly valued optically active trifluoromethylated epoxides **11c** (Scheme 2.25(c)).

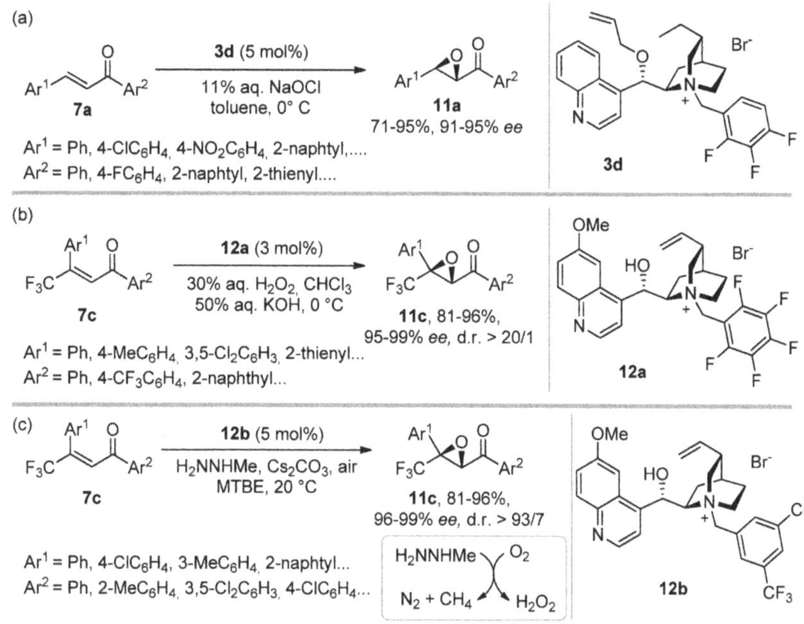

(a)

Ar^1 = Ph, 4-ClC$_6$H$_4$, 4-NO$_2$C$_6$H$_4$, 2-naphtyl,....
Ar^2 = Ph, 4-FC$_6$H$_4$, 2-naphtyl, 2-thienyl....

3d (5 mol%)

11% aq. NaOCl
toluene, 0° C

11a
71-95%, 91-95% ee

3d

(b)

Ar^1 = Ph, 4-MeC$_6$H$_4$, 3,5-Cl$_2$C$_6$H$_3$, 2-thienyl...
Ar^2 = Ph, 4-CF$_3$C$_6$H$_4$, 2-naphthyl...

12a (3 mol%)

30% aq. H$_2$O$_2$, CHCl$_3$
50% aq. KOH, 0 °C

11c, 81-96%,
95-99% ee, d.r. > 20/1

12a

(c)

Ar^1 = Ph, 4-ClC$_6$H$_4$, 3-MeC$_6$H$_4$, 2-naphtyl...
Ar^2 = Ph, 2-MeC$_6$H$_4$, 3,5-Cl$_2$C$_6$H$_3$, 4-ClC$_6$H$_4$...

12b (5 mol%)

H$_2$NNHMe, Cs$_2$CO$_3$, air
MTBE, 20 °C

11c, 81-96%,
96-99% ee, d.r. > 93/7

H$_2$NNHMe $\diagdown$ O$_2$

N$_2$ + CH$_4$ $\diagup\diagdown$ H$_2$O$_2$

12b

Scheme 2.25.

Last but not least, one should also mention the excellent performances of the axially chiral ammonium bromide **15b** developed by Maruoka *et al.* [OOI 04d]. As indicated in Scheme 2.26, PT catalyst **15b** displays a broad-substrate scope allowing not only the highly enantioselective epoxidation of chalcones **7a** but also the epoxidation of more challenging substrates such as enones bearing alkyl substituents at both the carbonyl carbon and double bond, including indanone and tetralone benzylidene derivatives **7d**. In all cases, epoxy ketones **11a** and **11d** were obtained in high yields (80–99%) and level of asymmetric inductions (89–99% *ee*), possibly making **15b** as one of the most efficient chiral ammonium PT catalyst described to date. The design of PT catalyst **15b** is made up of an *atropos* (*aS*)-binaphtyl core connected to a *tropos* biphenyl unit that would preferentially adopt an *aR* configuration at room temperature. In addition to these important stereochemical considerations, the presence of hydroxymethyl groups at the 3,3'-positions of the biphenyl unit appeared essential to ensure such efficiency, as shown by the low performances of PT catalyst **15c** (X = H) during the epoxidation of chacone **7a** (3%, 46% *ee*). It is assumed that the diphenylmethanol unit, far from being a simple structural element of the chiral pocket, would also plays a major role in the recognition and activation of the prochiral enone **7** by way of hydrogen bond interactions.

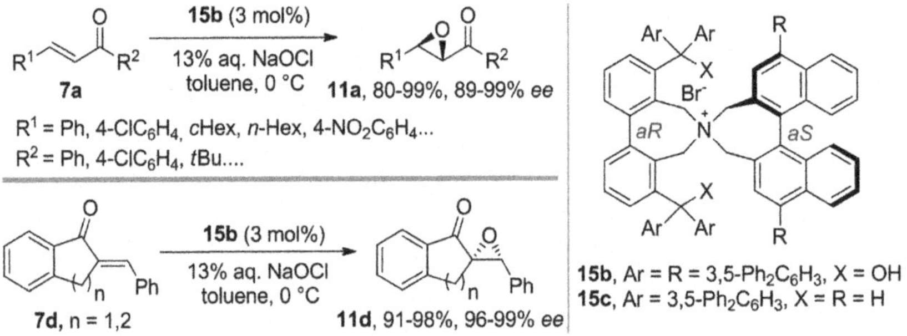

15b (3 mol%)

7a

13% aq. NaOCl
toluene, 0 °C

11a, 80-99%, 89-99% ee

R^1 = Ph, 4-ClC$_6$H$_4$, cHex, n-Hex, 4-NO$_2$C$_6$H$_4$...
R^2 = Ph, 4-ClC$_6$H$_4$, tBu....

7d, n = 1,2

15b (3 mol%)

13% aq. NaOCl
toluene, 0 °C

11d, 91-98%, 96-99% ee

15b, Ar = R = 3,5-Ph$_2$C$_6$H$_3$, X = OH
15c, Ar = 3,5-Ph$_2$C$_6$H$_3$, X = R = H

Scheme 2.26.

Enantioselective α-hydroxylation of ketones **1a** represents a challenging transformation giving rise to important chiral synthetic intermediates, namely tertiary α-hydroxyl carbonyl compounds **56** (Scheme 2.27). In 1988, Shioiri *et al.* reported the first enantioselective α-hydroxylation of ketones **56a** conducted under PTC conditions with molecular oxygen (Scheme 2.27(a), conditions A) [MAS 88]. The procedure included the use of triethyl phosphite to reduce the hydroperoxide intermediate arising from the reaction between the chiral ammonium enolate and molecular oxygen. In the presence of PT catalyst **3a**, various α-alkylated tetralones **56a** could be α-hydroxylated in good yields (87–98%) and satisfactory *ees* ranging in most cases from 70 to 77%. Despite this first promising result, it was only in early 2010 that a renewed interest in the α-hydroxylation of ketones mediated by chiral quaternary ammonium salts has emerged in the literature. In 2012, Meng *et al.* reported the α-hydroxylation of 1-indanone-derived β-oxo esters **1a** by means of cumyl hydroperoxide as oxidant (Scheme 2.27(a), conditions B) [YAO 12]. Mainly in indanone series, the use of the cinchoninium PT-catalyst **3c** equipped with two bulky groups on the oxygen atom at C-9 position (1-adamantoyl) and at the quinuclidine nitrogen (methylene-9-anthracenyl) was allowed to reach good asymmetric inductions (58–90% *ees*) and fair to good yields (45–90%) for a large variety of β-oxo esters **56a**. In the same year, Gao *et al.* investigated the photoxygenation of 1-indanone-derived β-esters **1a** by PTC using molecular oxygen or air (Scheme 2.27(a), conditions C) [LIA 12]. From a mechanistic point of view, irradiation by a light source is believed to promote the incorporation of active singlet oxygen to the chiral ammonium enolate so as to generate a α-hydroperoxide ketone intermediate, which would subsequently react, according to a disproportion process, with a second chiral ammonium enolate species and ultimately conduct to the desired α-hydroxy ketone **56a**. Probing the reaction scope with the simple cinchoninium PT catalyst **3a** revealed that the reaction performed well with a range of β-oxo esters **1a** in indanone series (*ees* up to 75% and 81–93% yields) under mild conditions. This approach provides the main advantage of using air as a cheap and environmentally benign oxidant, without requiring the use of an extra reducing agent such as triethyl phosphite (Scheme 2.27(a), condition C). As discussed previously in Scheme 2.11, Tan and colleagues also succeeded in the enantioselective oxygenation reaction of oxindole through a disproportion process by means of pentanidium catalyst **26** [YAN 12]. In 2008, Itoh *et al.* expanding the substrate scope of asymmetric α-hydroxylation to 2-oxindoles **1c** under very similar conditions to those reported by Shioiri *et al.* in 1988 [SAN 08, MAS 88]. By making use of cinchoninium PT catalyst **3a**, a range of 3-hydroxy-2-oxindoles **56b** were obtained in yields ranging from 91 to quantitative and with enantioselectivities ranging from 67 to 93% (Scheme 2.27(b)). Recently, an important breakthrough regarding the substrate scope was made by Zhao *et al.* who reported the α-hydroxylation of acyclic carbonyl compounds **1d** to give product **56c** by means of the readily available PT catalyst **6i** with fair to good yields (34–90%) and interesting levels of enantioselectivity (67–83% *ee*) [SIM 15]. The reaction was performed under aerobic conditions in the presence of 1,2-bis(diphenylphosphino)ethane (dppe) or triethyl phosphite as the reductant (Scheme 2.27(c)). Last but not least, the corresponding reaction conditions could also be successfully applied to several cyclic ketones (70–98% *ee*).

Beside electrophilic α-hydroxylation reaction of ketones, some attempts to use chiral quaternary ammonium salts in phase-transfer dihydroxylation of enones **7e** (having an aromatic and an aliphatic pendants) using potassium permanganate were initially made by Brown *et al.* (Scheme 2.28(a)) [BRO 01, BHU 02]. Although appealing levels of enantioinduction were obtained (**57a**, 67–80% *ee*), the required stoichiometric amounts of the Corey-type PT catalyst **6l** and the rather low conversions that were typically observed severely hamper the usefulness of this method. Interestingly, the same group reported an oxidative cyclization of 1,5-dieneones **7e'** conducted under similar conditions with as little as 5 mol% of **6l**, leading to the formation of tetrahydrofurane diols **58** (58–75% *ee*, 26–50%) while controlling the relative and absolute configuration of three newly created stereocenters in a single step (Scheme 2.28(b)) [BRO 01, BHU 02].

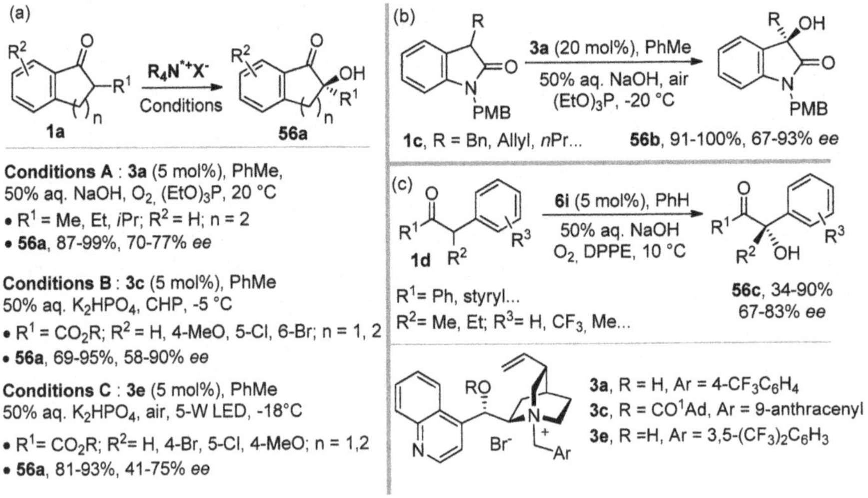

(a)
R4N*+X-
Conditions
1a
56a
Conditions A : 3a (5 mol%), PhMe,
50% aq. NaOH, O2, (EtO)3P, 20 °C
• R1 = Me, Et, iPr; R2 = H; n = 2
• 56a, 87-99%, 70-77% ee
Conditions B : 3c (5 mol%), PhMe
50% aq. K2HPO4, CHP, -5 °C
• R1 = CO2R; R2 = H, 4-MeO, 5-Cl, 6-Br; n = 1, 2
• 56a, 69-95%, 58-90% ee
Conditions C : 3e (5 mol%), PhMe
50% aq. K2HPO4, air, 5-W LED, -18°C
• R1= CO2R; R2= H, 4-Br, 5-Cl, 4-MeO; n = 1,2
• 56a, 81-93%, 41-75% ee
(b)
3a (20 mol%), PhMe
50% aq. NaOH, air
(EtO)3P, -20 °C
1c, R = Bn, Allyl, nPr...
56b, 91-100%, 67-93% ee
(c)
6i (5 mol%), PhH
50% aq. NaOH
O2, DPPE, 10 °C
1d
R1= Ph, styryl...
R2= Me, Et; R3= H, CF3, Me...
56c, 34-90%
67-83% ee
3a, R = H, Ar = 4-CF3C6H4
3c, R = CO1Ad, Ar = 9-anthracenyl
3e, R =H, Ar = 3,5-(CF3)2C6H3

Scheme 2.27.

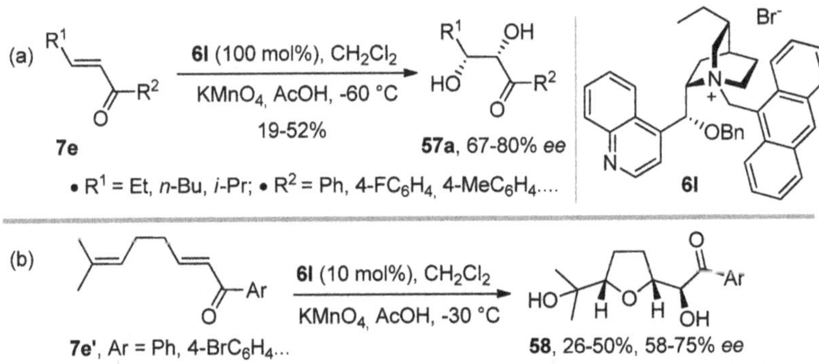

(a)
6l (100 mol%), CH2Cl2
KMnO4, AcOH, -60 °C
19-52%
7e
57a, 67-80% ee
• R1 = Et, n-Bu, i-Pr; • R2 = Ph, 4-FC6H4, 4-MeC6H4....
6l
(b)
6l (10 mol%), CH2Cl2
KMnO4, AcOH, -30 °C
7e', Ar = Ph, 4-BrC6H4...
58, 26-50%, 58-75% ee

Scheme 2.28.

Inspired by these pioneering works, Tan *et al.* recently developed the potassium permanganate mediated enantioselective oxidation of alkenes by using chiral dicationic bisguanidinium **60a–b** as catalysts (Scheme 2.29) [WAN 15b]. While α-aryl acrylates **7f** furnished diols **57b** in moderate yields (60–72%) and good asymmetric inductions (85–96% *ee*), the same reaction performed under acidic conditions with β-substituted α-aryl acrylates **7g** led to 2-hydroxy-3-oxocarboxylic esters **59** in good yields and excellent *ees* (84–96%). Strong experimental evidences suggest that both the chiral induction and the rate acceleration, which were observed, result from transition-state stabilization via ion pairing interactions between the chiral bisguanidinium cation **60** and the transcient enolate anion, while a mechanism based on PTC can be ruled out.

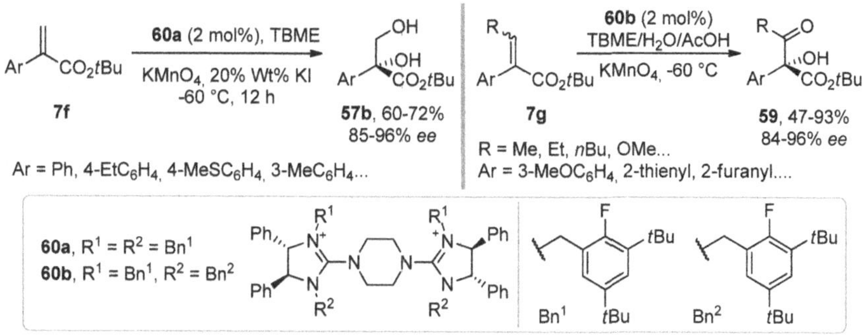

Scheme 2.29.

Although several papers deal with chiral quaternary ammonium salts-mediated enantioselective oxa-Michael reactions, one should mention the stereoselective preparation of trifluoromethyl-substituted 2-isoxazoline **61** reported by Shibata *et al.*, which features an oxa-Michael/cyclization/deshydratation a cascade reaction between hydroxylamine and trifluoromethylated enones **7c** (Scheme 2.30(a)) [MAT 10]. A catalyst screening revealed that quinidinium PT catalyst **12b** induced the highest asymmetric inductions (88–94% *ee*) with a wide range of trifluoromethylated enones **7c** and furnished 2-isoxazolines **61** in good yields (80–99%). Pleasingly, the authors noticed that the pseudoenantiomer of **12b** (the corresponding quininium salt) exhibited the same level of enantioselectivity, affording **61** with perfect inversion of the stereocontrol. A dramatic drop of the enantioselectivity was observed with *O*-methylated PT catalyst **12c** (34% *ee*), highlighting the crucial role of the hydroxyl group at C9 position in the stereochemical outcome of the reaction. An interesting example of chiral quaternary ammonium salts mediated cyclization of 2'-hydroxychalcones **7a** to flavones **62** was studied by Hintermann *et al.* (Scheme 2.30(b)) [HIN 12]. The reversibility of this intramolecular oxa-Michael reaction makes it tedious to find proper experimental conditions enabling both high conversion and enantioselectivity for this reaction. However, the use of the cinchoninium **3e**/ NaH as precatalyst system resulted in the formation of various flavones **62** with reasonable levels of enantioinduction (76–80% *ee*). An autocatalytic mechanism is put forward by the authors to account for the acquired experimental data. In an initial step, NaH would deprotonate chacone **7a** to generate an insoluble sodium chalconate, which after an ion metathesis process with cinchoninium chloride **3** would form a soluble chiral ammonium phenolate **7a'**. Intramolecular oxa-Michael addition within this chiral ion-pair **7a'** is expected to occur enantioselectively to produce enantioenriched enolate **62'**, which in last step would deprotonate of a new molecule of chalcone **7a** to release flavone **62** with concomitant regeneration of the chiral ion pair **7a'**.

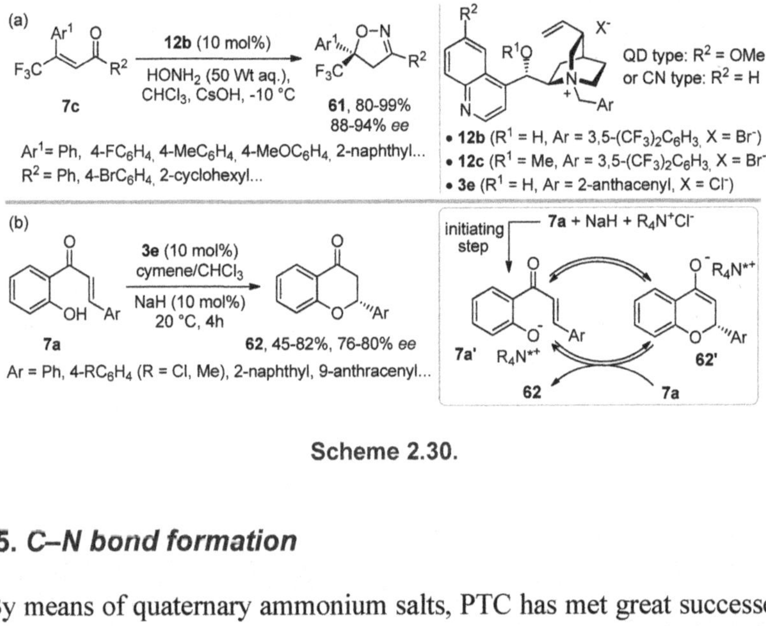

Scheme 2.30.

2.2.5. *C–N bond formation*

By means of quaternary ammonium salts, PTC has met great successes for the enantioselective construction of C–C and C–O bonds. Nonetheless, despite pioneering investigations [JUL 80], the efficient asymmetric construction of C–N bonds has only emerged recently [OOI 07b, SHI 13a]. With regard to a nucleophilic amine precursor, the efficient formation of the corresponding ammonium-amide ion pair is questionable both in terms of stability and reactivity, considering the significant basicity and nucleophilicity of a nitrogen-based anion. Second, as far as the useful aza-Michael reaction is concerned, the reversibility of the process may end-up in getting racemic mixtures. Accordingly, a subsequent cyclization event, which terminates the conjugated addition, is often sought after [HER 14]. Enantioselective C–N bond formation mediated by chiral quaternary ammonium salts was reported as early as 1996 by Prabhakar *et al.* during the aziridination of *t*-butyl acrylate **31g** (Scheme 2.31) [AIR 96, AIR 01]. The reaction conducted with *N*-pivaloyl-*N*-phenylhydroxylamine **63** and *t*-butyl acrylate **31g** in the presence of PT catalyst **6m** provided the corresponding aziridine **64** in 79% yield and a moderate 45% *ee*. The reaction performed from *O*-pivaloyl-*N*-phenylhydroxylamine furnished the aziridine with comparable yields and a similar level of enantioselection, giving strong evidence that the true aziridinating agent is likely the *N*-acyloxy anion **65**. Associated with the chiral ammonium cation, the intermediate **65** can promote the 1,4-addition reaction to give **66** undergoing an enantioselective domino cyclization as depicted in Scheme 2.31. In 2005, this aziridination approach was significantly improved by Murugan *et al.* (with up to 95% ee) because of a novel generation of *Cinchona*-derived PT catalyst [MUR 05]. Additionally, efficient procedures were also developed by means of other nitrogen donors such as *N*-chloro-*N*-sodiocarbamates or OMs-hydroxylamines [FIO 04, MIN 08, MUR 11].

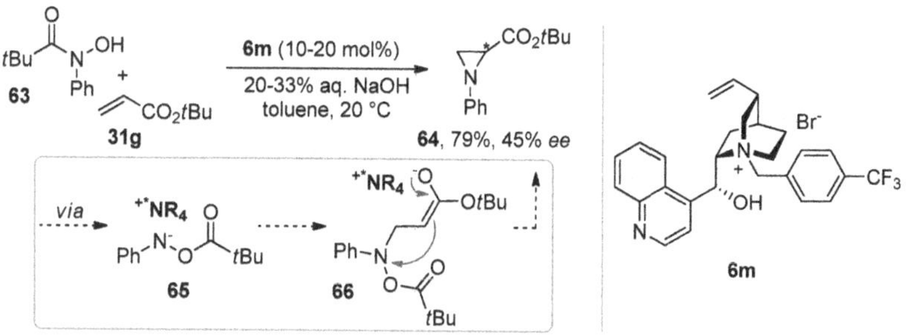

Scheme 2.31.

In 2013, Jørgensen *et al.* developed the first catalytic asymmetric diaziridination of *N*-tosyl aldimines **43c** with *N*-benzyl *O*-benzoyl hydroxylamine **67** under PTC affording optically active diaziridines **68** (Scheme 2.32) [LYK 13]. This class of strained three-membered ring aza-heterocycles is unique in two respects; (i) they not only disclose a dual reactivity due to the fact that they can be viewed as either hydrazines or aminals but also possess weak *N-N* bonds, making this class of compounds particularly appealing synthetic intermediates; (ii) the two nitrogen atoms are configurationally stables owing to both the ring strain and the lone pair repulsion, thereby conferring three stereogenic centers to this small ring. The best results were observed with cinchoninium PT catalyst **3f**, providing diaziridines **68** as single diastereoisomers in satisfactory yields (42–73%). As far as the aldimine **43c** scope is concerned, high to excellent *ees* were generally obtained with *meta-* and *para*-substituted aromatic imines **43c** (80–96% *ee*), whereas *ortho*-substituted aromatic imines and the aliphatic imine **43c** (R = cyclohexyl) led to rather disappointing results (33–41% *ee*). Maruoka also reported a PT-catalyzed asymmetric Neber rearrangement of oxime sulfonates in which the formation of a C–N bond intermediary led to enantioenriched azirines [OOI 02a].

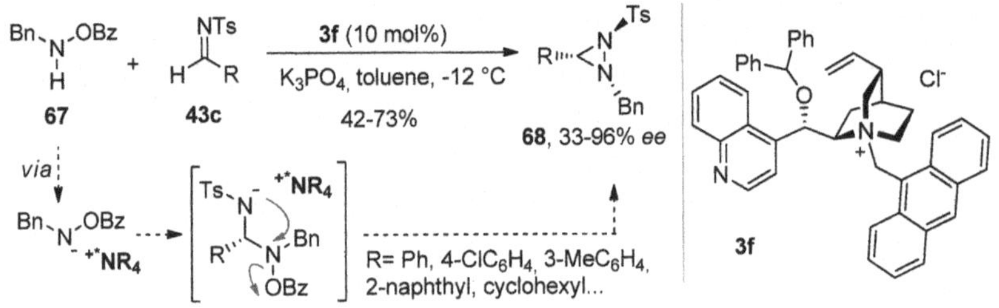

Scheme 2.32.

In 2008, Bandini and colleagues reported an efficient enantioselective synthesis of medicinally relevant pyrazino-indoles **70** (Scheme 2.33) [BAN 08, BAN 10]. Starting from acrylate **69**, the intramolecular character of the aza-Michael reaction prevented thermodynamic racemization. Interestingly, all attempts to perform the cyclization under chiral Brønsted base catalysis turned out to be much less efficient than the PT reaction performed by quaternary ammonium salts **6c**.

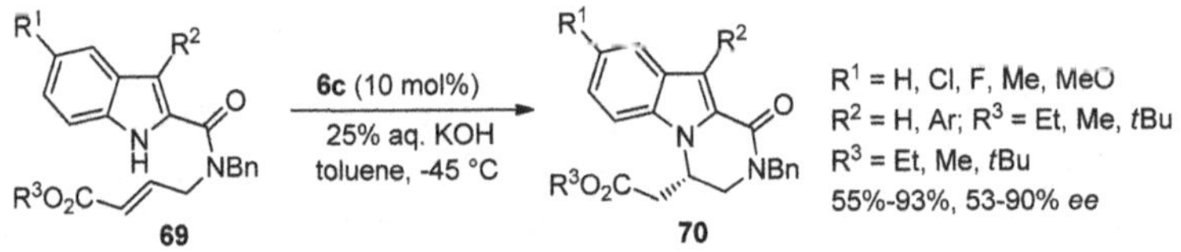

Scheme 2.33.

Deniau and Michon described a straightforward approach to isoindolinones **72** through the intramolecular aza-Michael reaction of **71** in the presence of Cs_2CO_3 (Scheme 2.34) [SAL 13, LEB 15]. A large screening of PT catalysts revealed that trimeric ammonium salt **3g** furnished the best enantiomeric excesses (up to 86% *ee*). The use of the *pseudo*-enantiomeric cinchonidinium (CD⁺) derived catalyst instead of the CN-catalyst **3g** led to a drop of *ee* (76–29% *ee* in otherwise identical reaction conditions), and highlighted that, thereby, the *pseudo*-enantiomeric effect is not always as straightforward as it is usually considered. As depicted in Scheme 2.34, the authors have shown that both the nature of the secondary amino group on the acrylamide moiety and the NHPh residue are key points for a successful asymmetric induction. In the vein of enantioselective PT-promoted intramolecular aza-Michael reactions, elegant applications were also reported in the context of the preparation of benzoindolizidines [GUO 15] and planar chiral aza-heterocyles [TOM 10], as well as in original for kinetic resolution processes of alkynes [MAI 11].

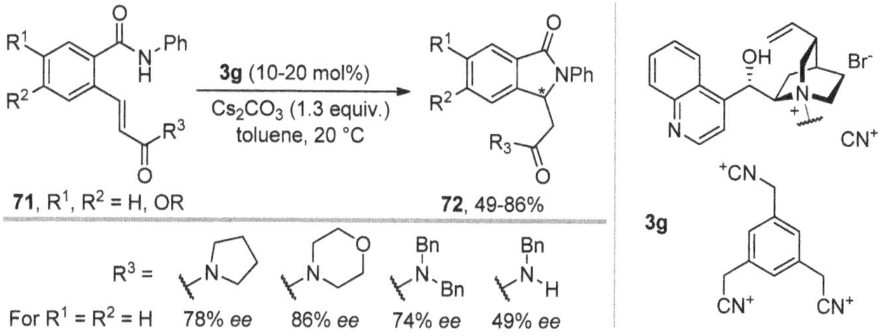

Scheme 2.34.

In 2010, Brière and co-workers demonstrated that solid–liquid PTC allowed the chemoselective deprotonation of *N*-Boc hydrazine **73** to promote a domino aza-Michael-cyclocondensation reaction with chalcones **7a** and, by this mean, developed a facile access to biorelevant 3,5-diaryl pyrazolines **74** (Scheme 2.35) [MAH 10, MAH 12]. The cyclocondensation event terminated the sequence by releasing a molecule of water and prevented any retro aza-Michael process to take place, securing thereby, a reaction under kinetic control. The use of the methoxy *ortho*-substituted benzyl quininium catalyst **9d** afforded the highest *ee* up to 92% in THF as solvent. Interestingly, a complete *pseudo*-enentiomeric effect was observed with quinidinium counterpart of catalyst **9d** allowing the construction of both enantiomer of pyrazolines **74**.

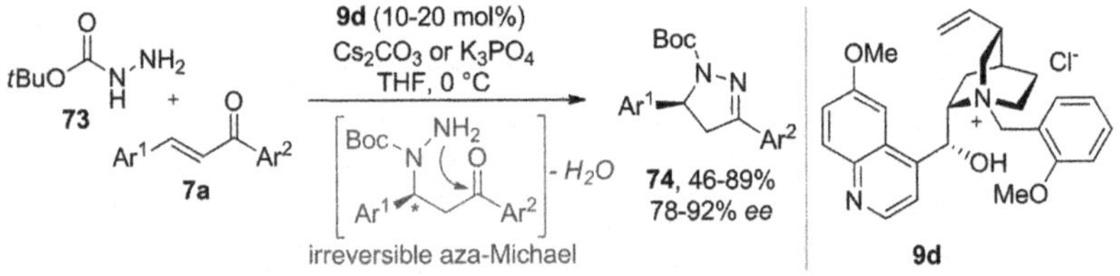

Scheme 2.35.

Considering the risk of a racemic pathway due to the base promoted retro aza-Michael process, the pure intermolecular 1,4-addition reaction of amines under PT catalysis remains challenging. Few successful examples made use of specific fluorinated-Michael acceptors or the innovative base-free conditions developed by Maruoka (see section 2.2.7) [WEI 12b, WAN 11]. Recently, Cho and Lee developed the aza-Michael reaction of azetidine **75** to enones **7e** flanked by linear alkyl chains at the β-position furnishing the corresponding adduct **76** with high *ee*s and only 1 mol% of the quinidine-based catalyst **12d** (Scheme 2.36) [LEE 15a]. Catalyst **12d** features a specific design with a OH functional group on the quinoline ring as prerequisite for getting high *ee*. The authors proposed the formation of a hydrogen bond between this acidic phenol residue and the nitro group of the deprotonated azetidine nucleophile **75** as an important interaction in the enantio-determining reaction step. The use of a catalytic amount of base (20 mol% of KOH) might involve a phase transfer initiated process, whereby the conjugated addition of the amide-anion leads to an enolate that deprotonates another amine **75**. On the other hand, initial formation of a betaine by the deprotonation of **12d** may also occurred (see section 2.3). Recently, Cho and colleagues reported a related PT-catalyzed intermolecular aza-Michael reaction of pyrazole to enones [LEE 15b].

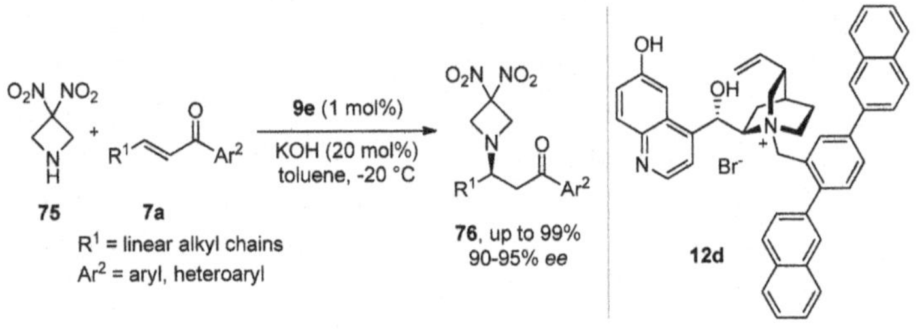

Scheme 2.36.

Besides the construction of stereogenic centers, it was also demonstrated, in few cases, that PTC may be capable of controlling other kind of chirality elements. In that context, Tomooka showed that PTC could direct the planar-chirality of nine-membered ring aza-heterocycles [TOM 10]. On the other hand, Maruoka succeeded in realizing the enantioselective synthesis of axially chiral *ortho*-iodoanilides **78** by means of an asymmetric alkylation of the amide moiety of **77** catalyzed by the *spiro*-ammonium **14c** (Scheme 2.37) [SHI 12]. The use of catalyst **14c** possessing extended aromatic substitutions (Ar groups) and *N*-hexyl chains was required for getting nice enantiomeric excesses. The authors beautifully succeeded in obtaining crystals of an ammonium-amide ion pair, which were well suited for X-ray diffraction analysis. On this basis, they proposed a model accounting for the good stereodifferentiation, which is observed when one of the ortho positions is substituted with a demanding iodine atom. Iodide derivatives furnished higher *ees* in comparison to other halogenated substrates (F > Cl > Br). Maruoka and collaborators later extended these concepts of PT atroposelective transformation to the kinetic resolution of 2-amino-1,1'-biaryl derivatives [SHI 13c].

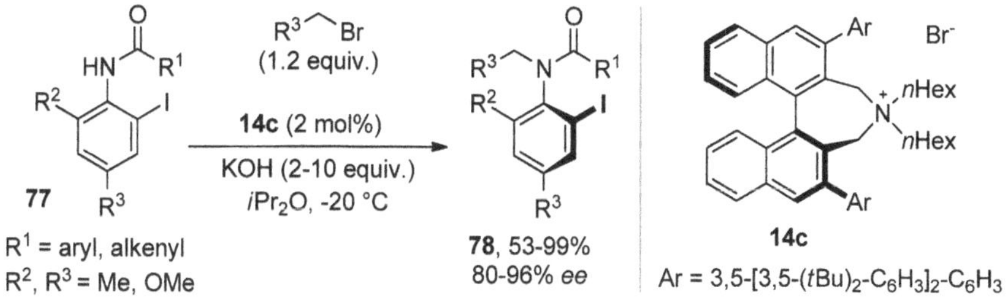

Scheme 2.37.

In addition to aza-nucleophiles, Jørgensen and colleagues reported the first attempts of using an aza-electrophile, namely di-*tert*-butylazodicarboxylate, for the PTC synthesis of 1,2,4-triazolines [MON 11].

2.2.6. *C–Y bond formation (Y = S, X...)*

Despite pioneering investigations at the earliest stages of the development of PTC, the efficacious asymmetric formation of C-heteroatom, except for C–O bonds, only recently re-emerged. In 1980, Juliá, Colonna and Wynberg independently reported the first examples of chiral ammonium salts promoted enantioselective kinetic resolution of secondary alkyl bromides and conjugate addition with thiolates [JUL 80, COL 81]. In 1999, Wladislaw reported on the kinetic resolution of sulfoxides through a PT-catalyzed sulfanylation reaction [WLA 99, ROD 10].

In 2007, Hou demonstrated that a chiral ammonium-thiolate ion pair $R_4N^+PhS^-$ smoothly formed from thiophenol **80** upon solid-liquid PT conditions allowed the enantioselective desymmetrization of *meso-N*-sulfonyl aziridine **79** (Scheme 2.38) [LUO 07]. By means of *N*-benzyl cinchoninium catalyst **3b**, the ring opening reaction occurred with up to 73% *ee* to give the β-sulfanylamine **81**.

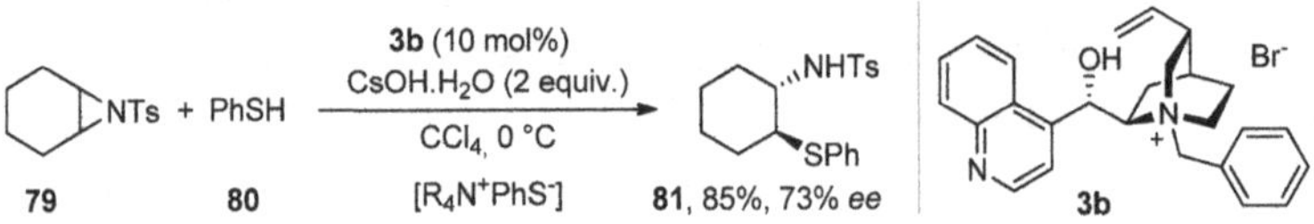

Scheme 2.38.

In 2011, Perrio and collaborators developed an innovative synthesis of enantioenriched sulfoxides **84** based on the enantioselective alkylation of sulfenate anion-ammonium ion-pairs **83** (Scheme 2.39) [GEL 11, GEL 13]. These ion pairs were easily generated *in situ* by a base-promoted retro thia-Michael fragmentation of precursors **82**. Importantly, the alkylation reaction with alkyl halides **27** mainly took place at the sulfur atom despite the ambident nucleophilicity of the sulfenate anion, which could have competitively led to *O*-functionalization products. Perrio and colleagues obtained their best (58%) starting from sulfinyl-sulfone precursors (EWG = SO₂Ph) **82** and methyl iodide **27** by means of the *N*-anthracenyl methylene cinchonidinium catalyst **6c**. In 2014, following Perrio's strategy, Kee and Tan made use of their novel pentanidium PT catalysts **26b** and achieved the synthesis of benzyl sulfoxides **84** ($R^1 = R^2 = $ Ar) from ester-sulfone precursors **82** (EWG = CO₂Me) [ZON 14]. This sequence turned to be very efficient for the synthesis of heteroarylsulfoxides **84** ($Ar^1 = $ Hetoaryl) and required 1 mol% of catalyst **26b**.

In 2014, Armstrong and Smith reported on the enantioselective formation of atropoisomeric sulfanyl-substituted biaryl derivatives **86–87** because of the enantioselective nucleophilic aromatic substitution reaction catalyzed by the *N*-benzyl quininium salt **9a** (Scheme 2.40) [ARM 14]. The PT-catalyzed thiolate addition not only allowed an asymmetric desymmetrization of the di-halogenopyrimidine precursors **85** but also introduced a tandem kinetic resolution process in which the minor enantiomers **86–87** were consumed via a second aromatic substitution with PhSH. The reaction furnished good *ee*s and yields except for sterically hindered substrate **86** with $R^1 = i$Pr.

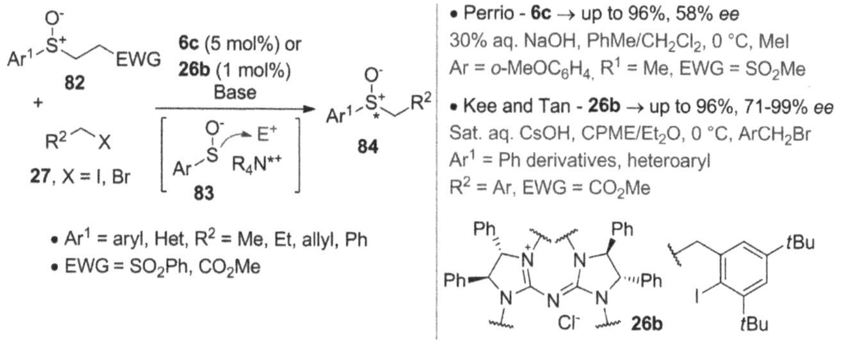

Scheme 2.39.

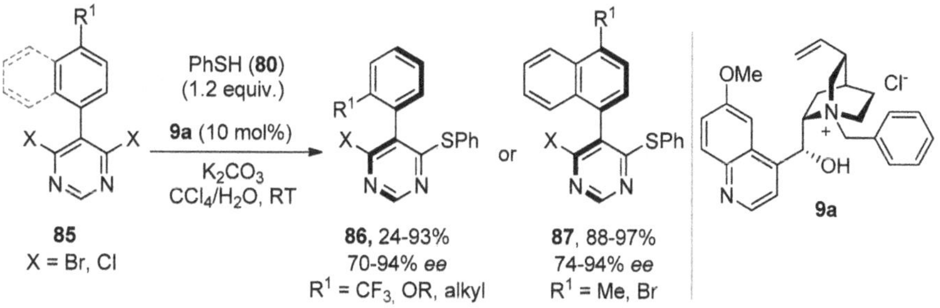

Scheme 2.40.

As far as electrophilic sulfanylation is concerned, Shen took advantage of his recently developed trifluoromethylthiolated iodine reagent **88** in order to perform the enantioselective functionalization of keto-esters **1a** (Scheme 2.41) [WAN 13c]. Upon liquid–liquid PT catalytic conditions in the presence of *N*-benzyl quininium salt **9a,Br**, the authors obtained the corresponding SCF₃-products **89** with *ee*s ranging from 66 to 96% for tetralone ($n = 2$) and benzosuberone derivatives ($n = 3$). Nevertheless, the *ee*s dropped down when five-membered ring indanone compounds **89** ($n = 1$) were used. This strategy provides an efficient approach to interesting products in medicinal chemistry with regard to the high lipophilicity and electron-withdrawing character of the SCF_3 functional group, which sometimes may enhance the pharmacokinetic properties of drugs.

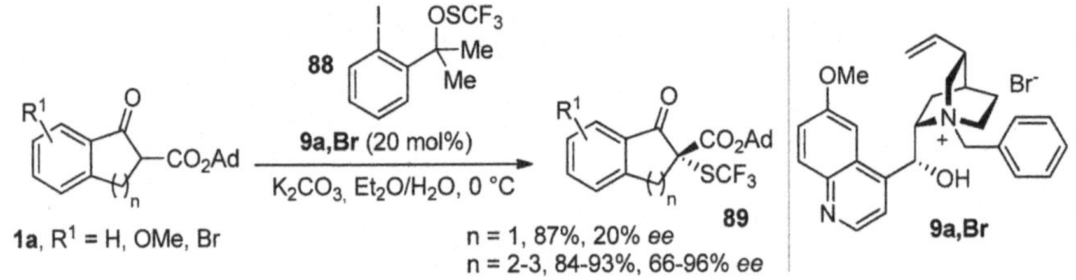

Scheme 2.41.

Recently, Zhenya Dai developed a PT-catalyzed sulfanylation reaction of glycine-imine by means of the reagents **90** furnishing products **91** with moderate to good *ee* (Scheme 2.42) [XIA 15]. To this end, the authors developed a novel hemiacetal ammonium salt **6n,** which is readily available in one step from the corresponding cinchonidine and the corresponding α-bromo acetophenone precursors.

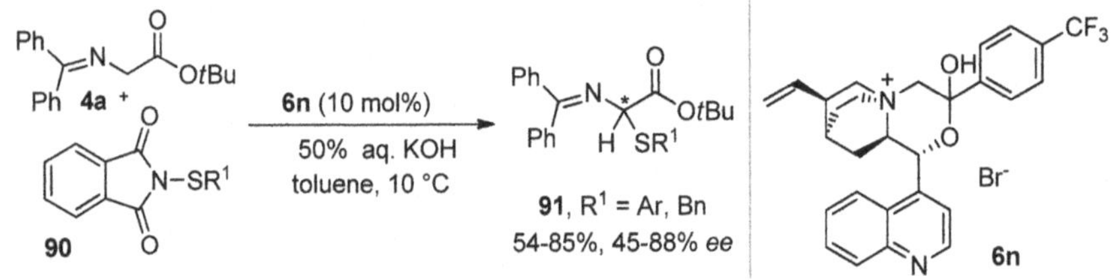

Scheme 2.42.

The asymmetric construction of other carbon heteroatom bonds (C–X) other than oxygen, amine and sulfur upon PTC has met with much less development thus far [OOI 07b, MAR 08b, SHI 13a]. For instance, Bernardi and Ricci developed a straightforward enantioselective entry to biorelevant α-amino phosphoric acid precursors **93** by means of a PT-catalyzed Pudovic-type reaction with catalyst **9e** (Scheme 2.43) [FIN 08]. Upon solid/liquid basic conditions, the α-amido sulfones **45** collapsed into the rather reactive *N*-carbamoyl imines **43b,** which underwent the nucleophilic attack of the dialkylphosphite **92** with high *ees*. This was claimed as the first PT-catalyzed hydrophosphination reaction, although the group of Lebel had also reported preliminary results concerning the PT-enantioselective alkylation of pro-chiral phosphine borane albeit in low *ee* [LEB 03].

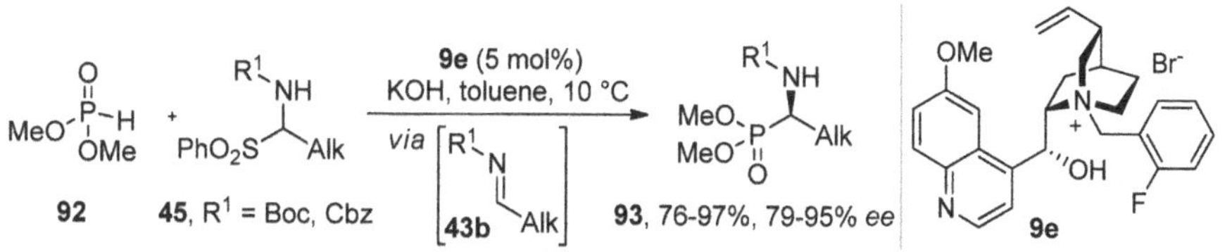

Scheme 2.43.

The construction of C–F bonds is a timely field of research with regard to the added value of fluorinated molecule in medicinal chemistry. Accordingly, a series of investigations tackled the enantioselective fluorination reaction of cyclic β-keto esters **1a** (Scheme 2.44). In 2002, Kim and Park reported the capability of *N*-fluoro benzenesulfonimide (**94**, NFSI) to achieve the introduction of a fluorine atom to 1-indanone derivative **1a** with 69% *ee* (Scheme 2.44(a)) [KIM 02]. The reaction proceeded upon solid–liquid PTC in the presence of the newly developed ammonium catalyst **3h**. Maruoka, in 2010, succeeded in the fluorination reaction of β-keto esters **1a** with excellent yields and ee (96% *ee*) by making use of the bifunctional *spiro*-ammonium PT-catalyst **14d** (Scheme 2.44(b)) [WAN 10]. As previously observed, the presence of hydroxyl functional groups on catalyst **14d** is a prerequisite for getting high enantiomeric excess. Furthermore, these conditions were successfully extended to other cyclic β-keto esters analogues of **1a** [WAN 10]. Interestingly, Gilmour and co-workers reported on similar fluorination conditions but making use of an original *N*-benzyl cinchoninium catalyst having a fluorine atom in place of alcohol at C9 [TAN 12]. Lu and colleagues recently demonstrated that the sterically hindered PT-catalyst **3c**, previously developed by Jørgensen, turned out to be an efficient promoter of the asymmetric fluorination reaction of β-keto esters **1a** with 90% *ee* (Scheme 2.44(c)) [LUO 13]. Interestingly, these PT conditions could be implemented with enantioselective chlorination reaction by means of *N*-chlorosuccinimide reagent [LUO 13]. In line with the new generation of bifunctional thiourea-containing PT-catalyst **20–21**, Lin and Duan have shown that the squaramide counterpart **96** was able to promote the asymmetric introduction of a fluorine atom to **1a** with 64% *ee* (Scheme 2.44(d)) [WAN 15a].

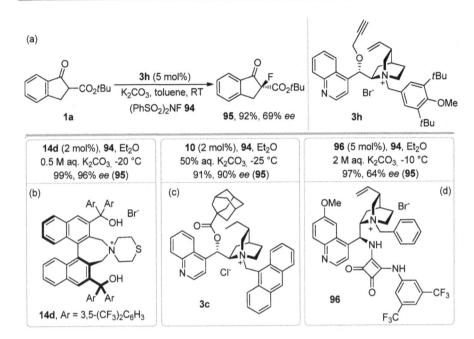

Scheme 2.44.

2.2.7. New developments in PTC

PTC has been driven for decades by the deprotonation of acidic nucleophiles by means of mineral bases before the formation of the chiral quaternary ammonium ion pair. In 2009, Maruoka et al. reported the base-free Michael addition reaction of oxindole **1c** to β-nitro-styrene **97** in the presence of the PT catalyst **14c** (Scheme 2.45) [HE 09]. The corresponding Michael-adduct **98** was obtained under very mild liquid–liquid water-rich biphasic conditions with high 90% enantiomeric excess and 93/7 distereomeric ratio. With model achiral ammonium salts (Scheme 2.43(b)), it was demonstrated that only lipophilic R_4N^+ salts with octyl chains led to good reaction rates securing a complete transformation in 2 h [SHI 14b, SHI 14c]. Interestingly, despite the catalysts **14e**, having alcohol functional groups (R^1 = H), furnished the corresponding Michael adducts with 90% ee, the methoxy homologue (**14f**, R^1 = Me) led to low 6% ee. This specificity was attributed to the importance of hydrogen bonding between the catalyst and the deprotononated-nucleophile **1c'**. As far as the mechanism is concerned (Scheme 2.45), the first step consists of the formation of enolate

1c' from **1c**. This equilibrated step was driven by the lipophilic ammonium salt favoring the formation of the ion pair **1c'** into the organic phase, while the liberated HBr is solvated in water phase. Accordingly, merely no reaction took place by running the reaction in homogenous conditions in toluene. Then, the enantioselective 1,4-addition reaction takes place to give the adduct **98'**, which is rapidly protonated by HBr that is in equilibrium between water and toluene phases. Thereby, the final product **98** is provided while the $R_4N^+Br^-$ catalyst **14e** is regenerated. This novel procedure is clearly an advance in terms of atom economy and improves the PTC-compatibility with base-sensitive substrates. Maruoka and colleagues successfully applied this strategy to various asymmetric reactions including, aza-Michael and aldol processes [WAN 11, SHI 14b, SHI 14c].

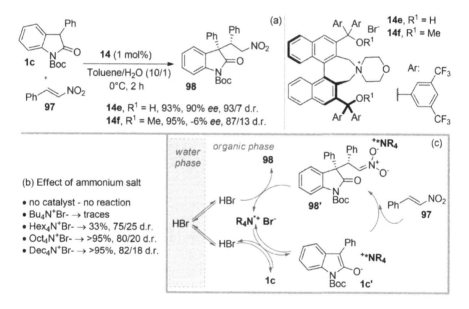

Scheme 2.45.

Fluorine-containing derivatives are an important class of compounds in medicinal chemistry but their syntheses require specific reaction conditions. For instance, β-ketoesters **1a** are reluctant to be perfluoroalkylated or trifluromethylated by iodide reagents **99** in regular PTC conditions (Scheme 2.46(a)) [WOZ 15]. Nonetheless, Melchiorre and colleagues developed an innovative photochemical-driven process allowing, upon visible light irradiation with white light-emitting diodes, allowing the synthesis of the corresponding fluorinated ketone **100** with good *ee* in the presence of 20 mol% of cinchoninium PT catalyst **3i**. The authors proposed the preformation of a photoactive electron donor–acceptor complexes **101** because of the interaction between the electron-rich ammonium enolate part and the electron-accepting perfluoroalkyl iodides **99** (Scheme 2.46(b)). Then, a single electron transfer process takes place upon irradiation to afford the reactive perfluoroalkyl radical (Rf•), which initiates a radical chain propagation catalytic pathway starting from the enantioselective C–C bond formation with enolate **102** upon the influence of the chiral ammonium cation moiety.

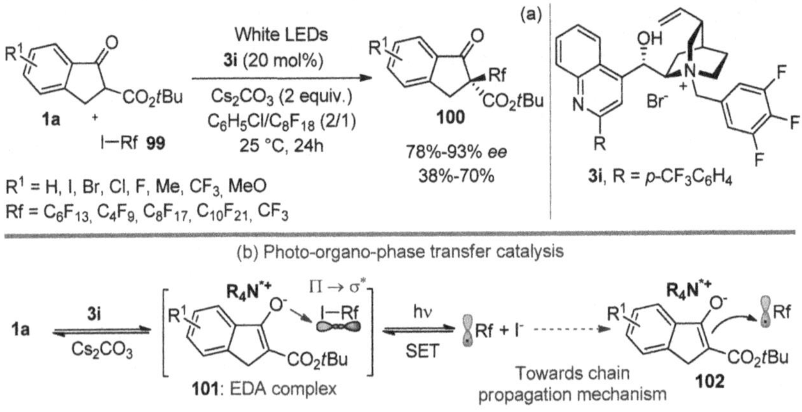

Scheme 2.46.

2.3. Cooperative ion pairs organocatalysts

As mentioned above, PT catalysts typically encompass a chemically inert counter-cation (i.e. halides in most cases). Following the development of new organocatalytic tools in the 2000s and the introduction of bifunctional catalysis, the concept that a counter-cation, acting as a Lewis and/or a Brønsted base, could cooperatively participate to the overall catalytic process was postulated (Scheme 2.47). As expressed by Levacher *et al.* in 2012, the cooperativity "leads to a synergistic outcome and implies that the two ions must participate cooperatively in the reaction mechanism either in terms of activation of reactive partners or by including chirality transfer" [BRI 12]. According to this principle, the involvement of such catalysts should allow reaching well-ordered transition states in which both reaction partners would be brought in close proximity and, accordingly, would operate with high efficiency and stereocontrol in a pseudo-intramolecular manner. By pushing further this principle, betaine catalysts have been designed by incorporating a covalent linkage between anionic and cationic part in order to increase the degree of organization in the transition state. In this section, we will only focus on the presynthesized quaternary ammonium catalysts. Accordingly, reactions involving the *in situ* formation of quaternary ammonium intermediates, as encounter during the acylation reaction catalyzed by DMAP, will not be considered.

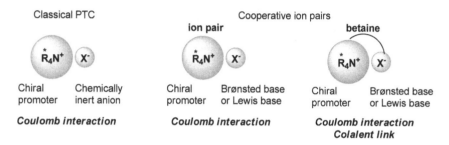

Scheme 2.47.

As depicted in Scheme 2.4, pioneering work was achieved by Wynberg and co-workers in the late 1970s by replacing classical halide by fluoride, thus giving rise to a Brønsted base catalyst that is able to achieve a Michael addition of nitromethane to chalcone **7a** in modest 23% *ee*. Nevertheless, this result paved the way for further developments in ammonium fluoride and hydrogen bifluoride catalysts based on either *Cinchona* alkaloid scaffolds or axially chiral dibenzazepine scaffolds (Scheme 2.48). Although being quite appealing, ammonium fluorides and, to a lesser extent, hydrogenbifluoride catalysts suffer from several limitations (the major one being their important hygroscopy), which led to the investigation of more efficient and more user-friendly ammonium salts catalysts. These studies notably resulted in the development of ammonium aryloxides first synthesized by Corey in the late 1990s for X-ray analyses purposes and then used as organocatalysts by Mukaiyama almost 10 years later [COR 97, TOZ 06a]. The construction of architectures in which the aryloxide part is covalently connected to the ammonium part provided the so-called betaine "an intramolecular ions pair" as demonstrated independently by Ooi, Gong and Levacher [URA 08, ZHA 12, CLA 13a]. In addition, few examples dealing with the use of acetates [PEN 07, PAT 09], hydride or hydroxide as counter-cations have been studied [HU 09, BLU 01].

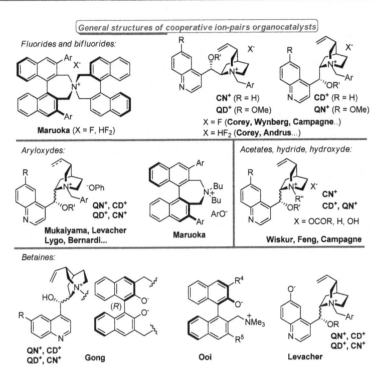

Scheme 2.48.

Before providing the reader with representative examples using those catalysts, we first intend to give more detailed information concerning the different reaction mechanisms typically at stake in ion paired quaternary ammonium salt organocatalysis.

2.3.1. Mechanisms of action

The replacement of inert halides in classical PT catalysts by fluorides, aryloxides, acetates, hydride and hydroxyde has important consequences on the mechanism of a given reaction. The most important relies on the fact that counter-cations are no longer chemically inert, and thus interact with at least one of the reaction partners, thus providing the opportunity to obtain a bifunctional mode of activation (Scheme 2.49). The chiral quaternary ammonium acts as a chiral inducer and is also able to activate the electrophile by stabilizing the development of the negative charge onto the addition product. The anionic part of the catalyst activates either a silyl or a protic nucleophile depending upon its Lewis or Brønsted base behavior, respectively. Such a dual activation within the ion pair catalysts results in the formation of supramolecular chiral entities and/or well-organized transition states, thus allowing to reach high level of stereoselectivity in several chemical transformations. As mentioned above, betaines could render the pseudo-intramolecular addition reactions because of the extra covalent link between anionic and cationic parts. Consequently, one can distinguish two types of modes of activation depending upon the intrinsic nature of the counter-cation.

The type 1 mechanism relies on the activation of a silylated pro- nucleophile ($NuSiR_3$) or a protic form (NuH) because of the Lewis (Scheme 2.49, Type 1a) or Brønsted (Scheme 2.49, Type 1b) base behavior of the counter-cation (X = F, HF_2, ArO or AcO), respectively. In the case of silyl nucleophiles, the first step consists of the activation of the silyl moiety thanks to the Lewis base properties of the counter-cation of the catalyst to provide a chiral hypervalent silicate that could transfer the nucleophile to an electrophile. The latter would be activated by the ammonium part of the catalyst through low-energy interactions (electrostatic or H-bonding). Regarding the protic nucleophiles, after a deprotonation step, a new chiral ion pair ($R_4N^+Nu^-$) possessing an acidic pendant (XH) is generated. The nucleophile adds to the electrophile that is activated through H-bonding. Finally, the ionic addition product is protonated by HX in order to provide the addition product with concomitant regeneration of the catalyst. Both secondary interactions (H-bonding or hypervalent silicate species) are also beneficial for the improvement of the organization of transition states and thus for the stereoselectivity of the reaction. Such mechanisms are well suited for 1,2 or 1,4-addition reactions, enantioselective protonation of silyl enolates or kinetic resolution of alcohol.

The second type of activation mode consists of activating the electrophile partner of the reaction and is encountered for ammonium betaine type catalysts (Scheme 2.49, type 2). This particular mode of activation is well suited for Steglich-type rearrangement. In this case, the phenolate moiety of the betaine catalyst adds to electrophilic center (E = Ac) of an achiral reagents (Nu-E), meanwhile generating the nucleophilic ammonium ion-pair ($R_4N^+Nu^-$). Then, the nucleophile undergoes an enantioselective pseudo-intramolecular acyl-migration upon the influence of the chiral ammonium moiety.

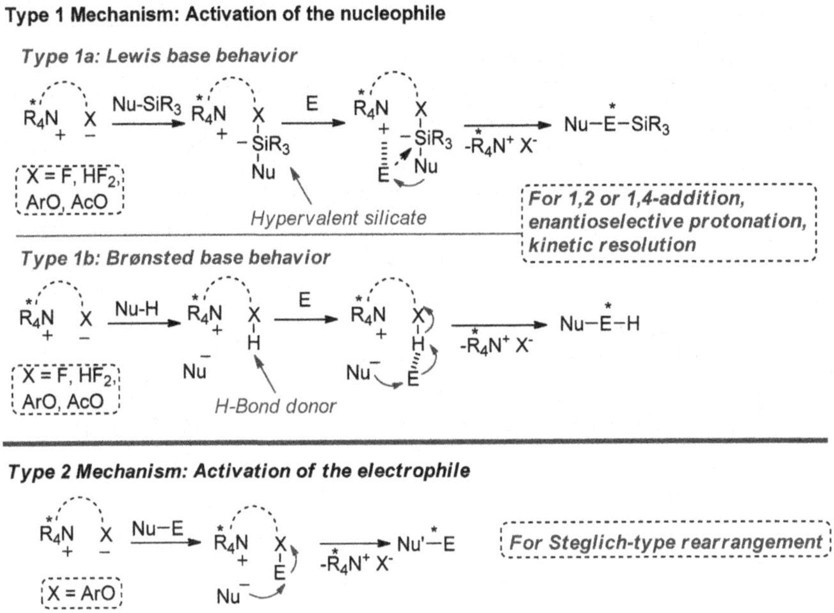

Type 1 Mechanism: Activation of the nucleophile

Type 1a: Lewis base behavior

Hypervalent silicate

For 1,2 or 1,4-addition, enantioselective protonation, kinetic resolution

Type 1b: Brønsted base behavior

H-Bond donor

Type 2 Mechanism: Activation of the electrophile

For Steglich-type rearrangement

Scheme 2.49.

Based on the aforementioned classification, we will now provide the reader with significant examples of selected applications of cooperative chiral ion pair organocatalysis.

2.3.2. Reactions based upon type 1 mechanism

2.3.2.1. Ammonium ion pairs used as Lewis base organocatalysts (type 1a)

As mentioned above, chiral ammonium salts flanked by a counter-cation possessing a Lewis base behavior (i.e. fluoride, hydrogenbifluoride, aryloxides, betaines, acetates, hydroxide, hydride) have been used in the activation of silyl nucleophiles. From a historical point of view, the first organocatalyst to be reported for such a purpose was the chiral ammonium fluoride. Indeed, in 1993, Shiori *et al.* reported on the synthesis and the use of chiral cinchoninium fluoride **3b,F** as organocatalysts for a vinylogous aldol reaction giving up to 72% *ee* (Scheme 2.50) [AND 93]. Catalyst **3b,F** was obtained by ion metathesis from their parent bromides **3b,Br** according to four different methods (A–D). Only slight variations in both isolated yields and enantiomeric excesses were observed depending upon the mode of preparation of the catalyst. The authors used the catalyst prepared from method C to probe the reaction. Several silyl enol ethers derived from 2-Me tetralone **103** were investigated using benzaldehyde **39b** as reaction partner and provided the addition products **104a** as an *erythro*-selective mixture of diastereomers with fair level of selectivity (70/30–82/18). The nature of the substituents on the aromatic moiety of the tetralone ring (R^1 group) had only little effect on both stereoselectivities (40–64% d.e. 66–70% *ee*) and isolated yields (67–74%). Similar level of yield and enantioselectivity (62%, 62% *ee*) were obtained with a silyl enol ether derived from pinacolone (**103c**, R^2 = *t*Bu), whereas a silyl enol ether derived from acetophenone (**103b**, R^2 = Ph) resulted in the formation of the corresponding addition product **104b** with a significantly lower enantioselectivity (39% *ee*).

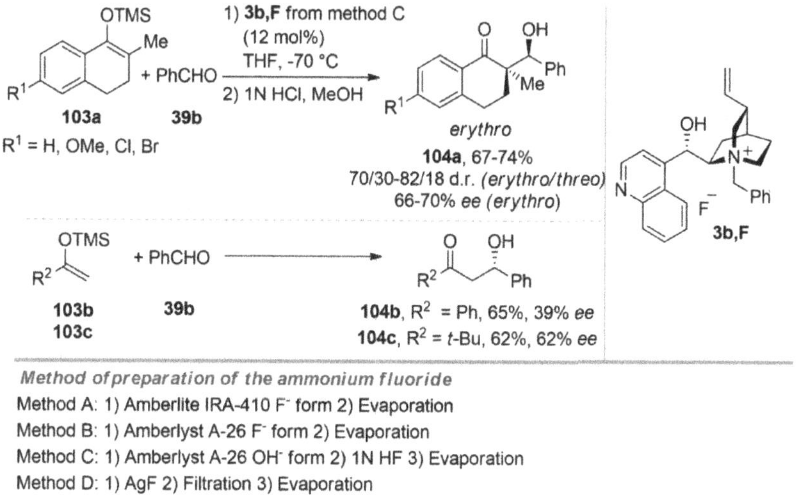

Method of preparation of the ammonium fluoride
Method A: 1) Amberlite IRA-410 F⁻ form 2) Evaporation
Method B: 1) Amberlyst A-26 F⁻ form 2) Evaporation
Method C: 1) Amberlyst A-26 OH⁻ form 2) 1N HF 3) Evaporation
Method D: 1) AgF 2) Filtration 3) Evaporation

Scheme 2.50.

Despite these interesting results, such catalysts suffered from a major drawback linked to their high hygroscopicity. As a result, in the following decades, this issue has attracted a lot of attention from the academic community that led researchers to propose two different solutions: (i) avoid the isolation of the catalysts by developing an *in situ* generation strategy and (ii) use a less hygroscopic salt having similar Lewis base properties to fluoride by using other anions.

An elegant example was reported by Shibata *et al.* during the trifluoromethylation of ketone **1** using the Ruppert–Prakash reagent **105** (TMSCF$_3$) (Scheme 2.51) [MIZ 07]. Indeed, starting from a chiral ammonium bromide and an achiral fluoride source, the chiral ammonium fluorides were generated *in situ* by ion metathesis. In order to elect the most straightforward and convenient source of fluoride, several additives were screened such as KF/2H$_2$O, TBAT, TBAF/H$_2$O and tetramethylammonium fluoride (TMAF). Among them, TMAF has brought a good balance in terms of yield and enantioselectivity for the naphthyl methyl ketone **1e**, which is used as model substrate. After an intensive screening of the conditions, a dimeric chiral biscinchoninium dibromide **3j** in combination with TMAF (10 mol% of each) in toluene/DCM (2/1) at a temperature ranging from –50 to –60°C was set as the best conditions. These conditions revealed to be quite general providing high level of enantio induction (**106a**, 70–94% *ee*) along with good isolated yields (59–96%) for a wide range of aromatic ketones **1d**. The aromatic ketones substituted by electron-withdrawing group **1d**, aliphatic ketones **1f** or naphtaldehyde **39c** exhibited a lower level of stereoselectivity. Nevertheless, this report represents an important breakthrough in comparison to the previous publications using isolated chiral ammonium fluorides that only provided the trifluoromethylated alcohol **106** resulting from the aromatic ketones in only up to 48% *ee* [ISE 94].

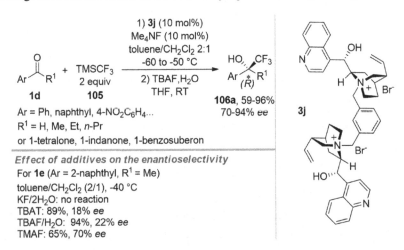

Scheme 2.51.

The other solution for solving the hygroscopicity issue associated with chiral ammonium fluoride is to replace the fluoride by other Lewis base anions such as bifluoride, aryloxide or acetate. Bifluoride anion was explored as an alternative to fluoride and the first report was by Corey *et al.* in 1999 which used a *Cinchona* alkaloids ammonium salt [HOR 99]. A few years later, Maruoka *et al.* extended this approach to their highly efficient structurally rigid C_2-symmetric chiral quaternary ammonium **13e** (Scheme 2.52) [OOI 03]. The resulting chiral ammonium bifluoride **13e,HF$_2$** was synthesized according to a slightly modified Shioiri's procedure (procedure C, Scheme 2.50) starting from the parent ammonium bromide **13e,Br**. These catalysts were successfully applied to the *anti*-selective 1,2-addition of silyl nitronate **107** to aromatic aldehydes **39d**. After subsequent desilylation, the resulting β-nitro alcohols **108** were obtained in high yields (70–94%), modest to high diastereomeric ratio (57/43 to 94/6) and excellent enantiomeric excesses (91–97%). Note that changing the nature of the silyl group from a Me$_3$Si or Et$_3$Si to bulky *t*-BuMe$_2$Si resulted in decreased yields, dia- and enantioselectivities. Interestingly, when switching to α,β-unsaturated aldehydes, highly chemoselective 1,4-addition (32/1) took place giving the silyl enol ethers of γ-nitro aldehydes that are stable enough to be obtained by flash chromatography purification in high yields (>87%) and excellent enantioselectivities (90–98% *ee*) for the major *anti* adduct (*anti/syn*: 76/24–97/3). These results clearly highlight the effectiveness of C_2-symmetrical chiral ammonium hydrogenbifluorides as Lewis base organocatalysts for the activation of silyl nitronate.

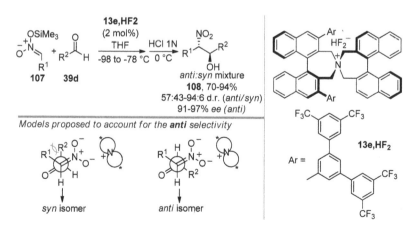

Scheme 2.52.

Following the same philosophy, Mukaiyama *et al.* evaluated the potential of chiral ammonium phenoxides derived from *Cinchona* alkaloids as Lewis base organocatalysts in asymmetric transformations [TOZ 06a, TOZ 06b, NAG 06, MUK 06, TOZ 07, NAG 07a, MUK 07]. Chiral ammonium phenoxides were obtained from the parent ammonium halides by the treatment with an ion exchange resin (Amberlyst A26 OH⁻ form) to afford the corresponding ammonium hydroxide, which is then quenched by phenol (1.0–2.0 equiv.). After removal of water by co-evaporation with benzene and crystallization, chiral ammonium phenoxides derived from *Cinchona* alkaloids can be isolated. As an example of the use of chiral ammonium phenoxides and for comparison purpose, one can cite the trifluoromethylation of aromatic ketones **1d** by the Ruppert–Prakash reagent **105** already discussed above using ammonium fluoride as catalysts (see Scheme 2.51). Indeed, Mukaiyama *et al.* reported in 2007 on the enantioselective addition of TMSCF₃ **105** to several aromatic ketones **1d** using cinchonidine-derived quaternary ammonium phenoxides **6,PhO** (Scheme 2.53) [NAG 07b]. The nature of the Ar substituent of the catalyst plays a crucial role in the stereocontrol. Enantioselectivities ranging from 8% (**6o,PhO**; Ar = 2,6-F₂C₆H₃) to 62% (**6p,PhO**; Ar = 3,5-[3,5-(CF₃)₂C₆H₃]₂C₆H₃) were obtained for the reaction of 3-nitroacetophenone **1g**. Catalyst **6p,PhO** was then chosen for studying the scope of the reaction using toluene/DCM (7/3) as the solvent system at −78°C. Although almost quantitative yields (addition product **106b** isolated as a silyl ether) were obtained independently of the substitution pattern of the aromatic ring of the aromatic ketones, modest enantioselectivities ranging from 46 to 77% were observed. The *S* absolute configuration of the addition product was ascertained by X-ray crystallographic analysis after conversion into the corresponding carboxamide **109**. For this particular reaction, the catalytic system reported by Shibata *et al.* (chiral ammonium bromide/TMAF) provided higher *ee* along with the advantage of avoiding the isolation of the active chiral catalyst.

Following the pioneer work of Mukaiyama, several groups have studied the *in situ* generation of chiral ammonium aryloxides by ion metathesis and successfully applied these conditions to several reactions such as trifluoromethylation of aldehydes [ZHA 07], imines [BER 12], direct vinylogous aldol reaction of (*5H*)-furan-2-one derivatives [CLA 13b], desymmetrization of 4-substituted cyclohexanones and the enantioselective protonation of silyl enol ethers [CLA 13a, CLA 13c]. The latter reaction is of particular interest when it comes to underline the positive impact of a betaine catalyst versus the corresponding simple ammonium aryloxide. Indeed, enantioselective protonation of silyl enol ethers derived from 2-substituted tetralones (**103a**, X=(CH$_2$)$_2$), indanones (**103d**, X=CH$_2$) and benzosuberones (**103e**, X=(CH$_2$)$_3$) was achieved by using chiral ammonium 4-MeOphenoxides derived from *Cinchona* alkaloids as a catalyst and a stoichiometric amount of 4-MeOphenol as an achiral external source of proton (Scheme 2.54). The quininium series **9** exhibited a higher level of stereoinduction than quinidinium **12**, cinchoninium **3** and cinchonidinium **6** structures in THF at –20°C using silyl enol ether of 2-Me tetralone **103a** as model substrate. More interesting is the comparison between ammonium aryloxide **9f,4-MeOPhO** and betaine **9g**, which exhibits the same quininium backbone but provided the enantioenriched ketone **1a** in significantly better enantioselectivity (12% vs. 51%, respectively). This result clearly demonstrates the advantage of using a betaine rather than a simple ammonium phenoxide. This superiority of the betaine catalyst over classical ammonium aryloxide could be explained from a mechanistic point of view by considering a pseudo-intramolecular transfer of the proton in one of the postulated transition states **A**. The betaine **9g** was thus involved in the reaction with several silyl enol ethers **103** allowing to obtain the corresponding enantioenriched ketones **1a** in high yield (>80%) but modest level of enantioselectivity (up to 62% *ee*).

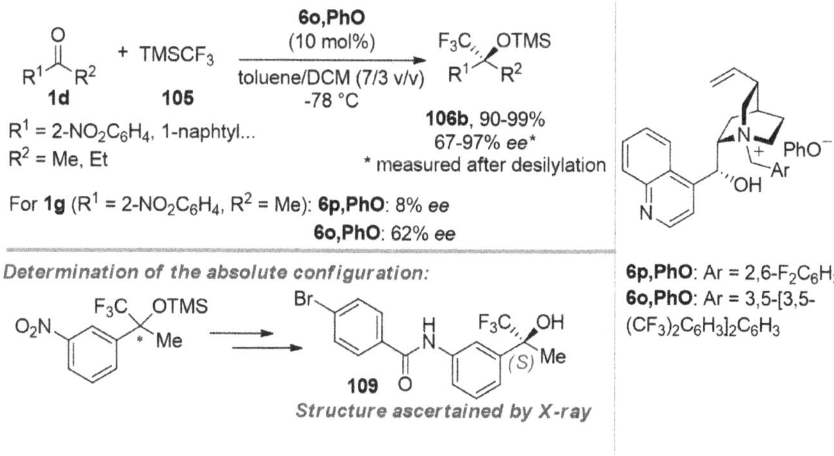

Scheme 2.53.

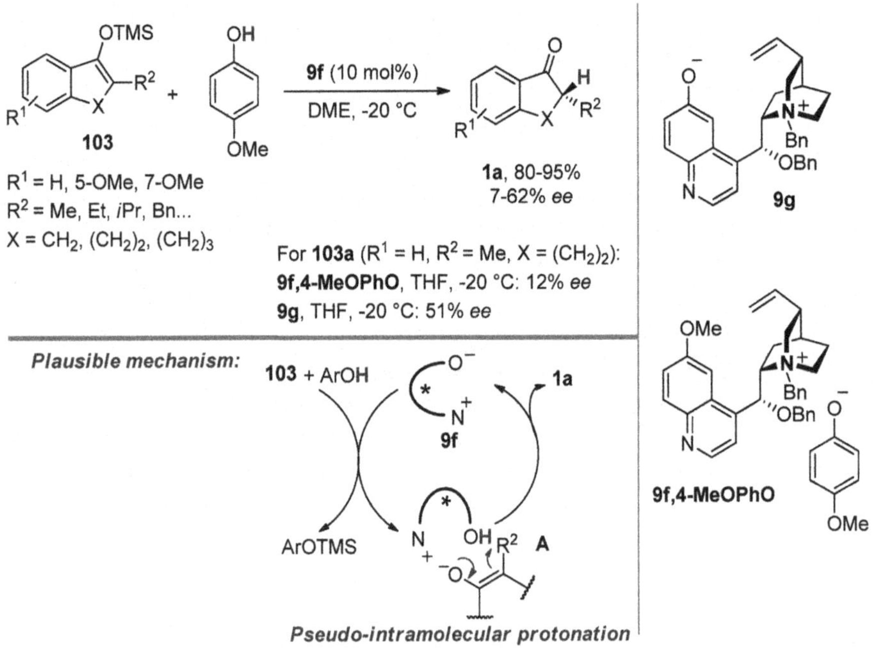

Scheme 2.54.

Lastly, although less developed, acetate counterion has also been associated with chiral quaternary ammonium salts [PEN 07, PAT 09]. By using such catalysts, Wiskur *et al.* published a Mukaiyama aldol reaction of silyl ketene acetals **110** to aromatic aldehydes **39d** catalyzed by *N*-Me cinchonidinium acetate **6q,AcO** (Scheme 2.55) [PAT 09]. Although this reaction is already well known in the literature, the authors revealed an original stereochemical aspect of this reaction. Indeed, the reaction is composed of two sequential reactions, namely a C–C bond formation and an *O*-silylation. A careful examination of the reaction shows that the *ee* changed over the course of the reaction. To account for this observation, the author proposed a mechanism based upon a racemic C–C bond-forming reaction followed by a kinetic resolution process occurring during the silylation step. From a mechanistic point of view, the first step would be the activation the silyl ketene acetal **110a** because of the Lewis base properties of the acetate anion. After the addition of the ketene acetal **110a** to the aldehyde **39d**, the corresponding racemic alkoxide is paired with a cinchonidinium cation **6,** thus forming diastereomeric salts. These salts then act as new catalysts and the enantioselectivity is supposed to arise from the difference of reactivity of these salts toward the silyl source (silyl ketene acetal or silyl acetate) during the silylation step. As a result, one aldol enantiomer is silylated **111-TMS** while the other remains under an alkoxyde form that is protonated upon treatment **111**. Although high global yields (>78%) and interesting *ee*s were obtained (68–86%), fair levels of selectivity factor (s factor) ranging from 1.8 to 3.7 were observed for the product **111** [KAG 88].

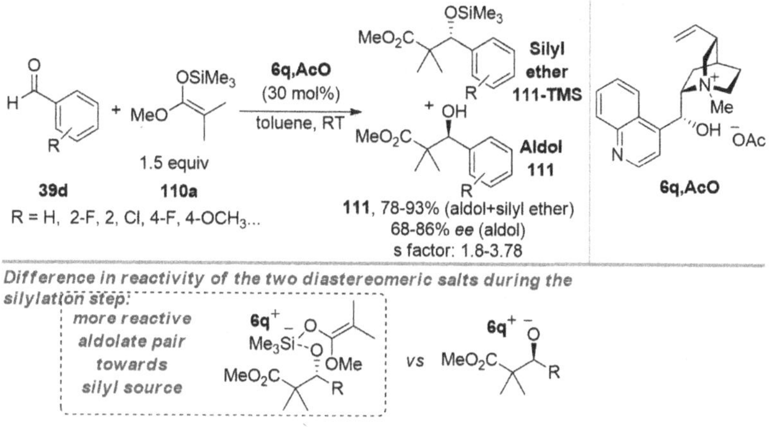

Scheme 2.55.

Finally, we would like to mention the use of two other anions that have been scarcely used as Lewis basic counter-cation of chiral ammonium, namely hydroxide and hydride. In 2001, Campagne *et al.* reported chiral quaternary ammonium hydroxide to be a Lewis base catalyst in the asymmetric vinylogous Mukaiyama aldol reaction (Scheme 2.56) [BLU 01]. Indeed, only one example is reported in the paper using *N*-Bn cichonidinium hydroxide **6a,OH**, thus furnishing the aldol product **112** in quantitative yield but with low level of enantioselectivity (< 30%). The corresponding cinchonidinium fluoride **6a,F** has also been evaluated providing almost the same yield and enantiomeric excess. Nevertheless, note that cinchonidinium hydroxide **6a,OH**, being an intermediate in the synthesis of the corresponding fluoride **6a,F**, its use as catalyst seems more appealing and would deserve to be extended to other reactions. Nevertheless, the stability of this highly basic species has to be taken into account.

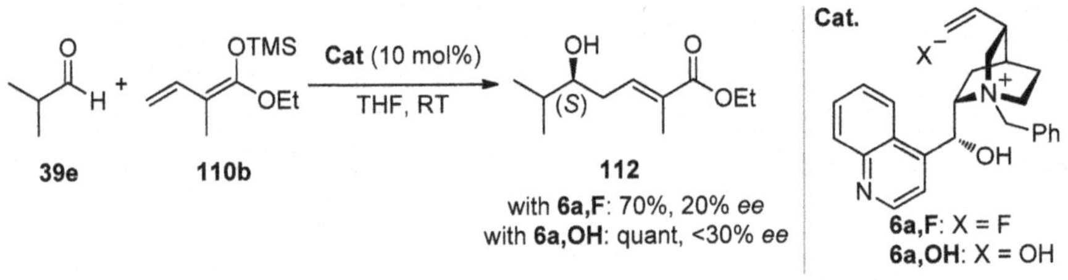

Scheme 2.56.

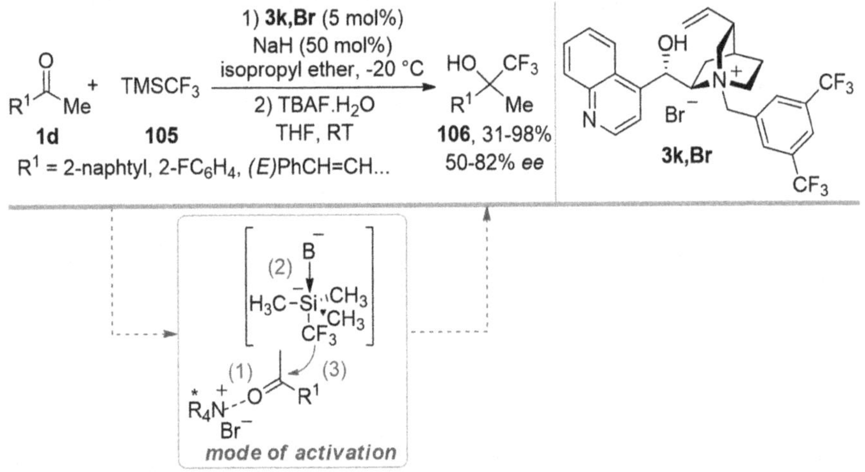

Scheme 2.57.

The other example deals with the use of chiral ammonium hydride as catalysts in the trifluoromethylation of aromatic ketone **1d** by the Ruppert–Prakash reagent **105** (Scheme 2.57) [HU 09]. By using a combination of chiral cinchoninium bromide **3k,Br** (5 mol%) and substoichiometric amount of sodium hydride (50 mol%), variable yields ranging from 31 to 98% along with fair to high enantiomeric excesses (50–82% ee) were obtained. Except for 1- or 2-naphtaldehyde **39c** (82% and 81% ee, respectively), the use of other substituted phenyl substituents or benzylideneacetone resulted in a decrease in the enantiomeric excesses (50–68% ee). To account for the stereoselectivity observed, the authors proposed a scenario in which (i) the carbonyl is first activated by the ammonium part of the catalyst allowing to discriminate the two enantiotopic faces of the ketone **1d**, and (ii) the TMSCF$_3$ **105** is activated by the Lewis base to form an hypervalent silicate intermediate that could (iii) transfer easily the trifluoromethyl anion to provide the addition product **106** and upon silylation regenerates the chiral ammonium cation flanked by the Lewis base.

2.3.2.2. Ammonium ion pairs used as Brønsted base organocatalysts (type 1b)

The use of a chiral ammonium flanked by a counter-cation with basic properties and thus able to activate protic nucleophiles is less developed in the literature compared to the activation of silyl nucleophiles discussed in the previous section. This is surprising if one considers that the initial development of the quaternary ammonium fluoride in 1978 was in the Henry reaction in which a fluoride played the role of a Brønsted base [COL 78]. Thus, except fluorides, all the other examples involved aryloxides. As we will see below, betaines have also been investigated as a Brønsted base catalyst.

Aryloxides are interesting from a structural point of view because the introduction of substituents on their aromatic ring can lead to an important variation in their pKa. For example, 4-MeOC$_6$H$_4$OH shows a pKa of 19, whereas 4-NO$_2$C$_6$H$_4$OH exhibits a lower pKa value of 11 (for a list of pKa, see: http://www.chem.wisc.edu/areas/reich/pkatable/). Thus, this high magnitude of pKa makes these aryloxides well suited for the design of finely tuned Brønsted basic counter-cations. The first application of chiral quaternary ammonium aryloxides was reported by Lygo *et al.* in 2009 and concerned the Michael addition of *tert*-butyl glycine imines **4a** to various Michael acceptors (Scheme 2.58) [LYG 09b]. The ion pair catalyst consisting of dihydrocinchonidinium **6l** flanked by an aryloxide (ArO = 2,4,6-Me$_3$C$_6$H$_2$O) was generated *in situ* by ion metathesis between the deprotonated mesitol (2,4,6-trimethylphenol + KOH) and the parent dihydrocinchonidinium bromide **6l,Br**. With this catalytic system, high yields (63–99%) and excellent enantiomeric excesses (89–98% *ee*) were obtained for a wide range of Michael acceptors (acrylates, alkyl vinyl ketones, acrylonitriles, phenyl vinyl sulfone). While vinyl phosphonates and acrylamides remained unreactive. From a mechanistic point of view, it is important to note that no reaction took place in the absence of mesitol (chiral ammonium halide + solid KOH) arguing in favor of **6l,ArO** as the active catalytic species. Another point that should be underlined is the influence of the phenol derivatives onto the enantioselectivity. Indeed, while simple phenol provided the 1,4-addition product in 81% *ee*, mesitol resulted in an increasing of the *ee* to 98%. Thus, the first step of the mechanism would be the *in situ* generation of **6l,ArO** by ion metathesis between the corresponding ammonium bromide and the aryloxide resulting from the deprotonation of mesitol by solid KOH. The rest of the mechanism seems to be more complicated than expected as the authors mentioned the fact that if solid KOH is removed from the reaction medium before addition of glycine imine and Michael acceptor, only low conversion is observed. Nevertheless, this protocol provides a useful complement to the traditional liquid–liquid PT conditions (aq. KOH, PTC) that have been reported to be efficient in the alkylation of imines **4a**.

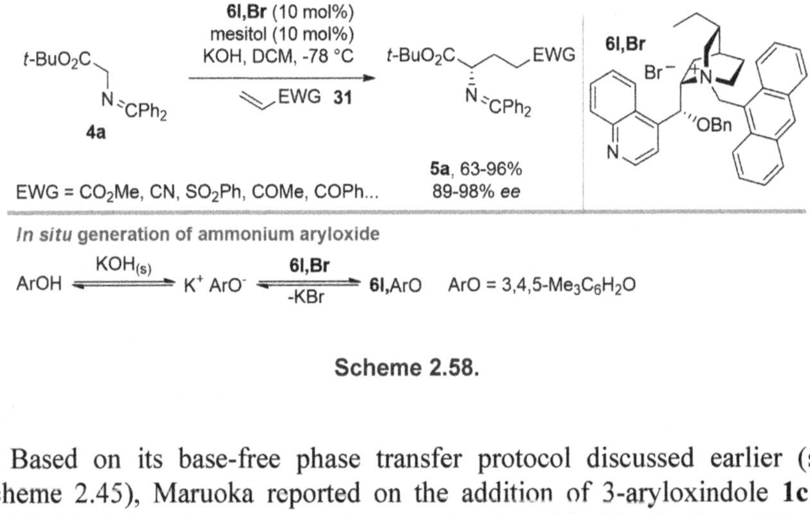

Scheme 2.58.

Based on its base-free phase transfer protocol discussed earlier (see Scheme 2.45), Maruoka reported on the addition of 3-aryloxindole **1c** to maleimide **113** in the presence of ammonium bromide **14a,Br** (1 mol%) and 3,5-dimethylphenol (5 mol%) (Scheme 2.59) [SHI 14a]. The corresponding addition products **114** were isolated in good yields (65–95%) and high diatereo- and enantioselectivities (91:9-99:1 d.r. and 80–93% *ee*). The ammonium aryloxide **14a,ArO** was put forward by the authors as being the truly active catalytic species. As a result, the mechanism would start by the formation of corresponding ammonium enolate **A** resulting from the displacement of the keto-enol equilibrium because of the bromide part of the PTC. This enolate would then add to the maleimide **113** providing the addition product under the form of a new enolate **B** that would be subsequently protonated by the phenol derivatives, thus delivering the addition product **114** with concomitant generation of an ammonium aryloxide **14a,ArO**. This latter ion pair would be the new catalytic species for the next catalytic cycles. In order to ascertain this mechanism, the authors have performed the reaction in the presence of 1 mol% of ammonium aryloxide **14a,ArO** previously isolated, thus furnishing the addition product in comparable level of stereoselection than the one obtained with the PTC/aryloxide system.

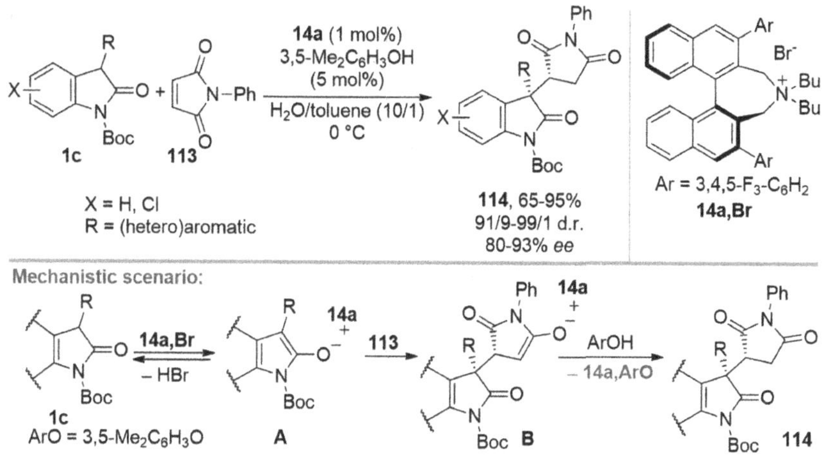

Scheme 2.59.

Ooi and colleagues also applied their betaine catalysts **115** as a Brønsted base in Mannich type reactions with several protic nucleophiles including nitroalkanes [URA 08, URA 10c, URA 15], β,β-disubstituted nitroolefins [URA 12b], α,β-disubstituted nitroolefins [OYA 15] and 2-benzyloxythiazol-5(*4H*)-ones [URA 10b]. In order to illustrate the efficiency of these classes of catalyst, we selected the recent publication of Ooi *et al.* dealing with the vinylogous addition of α,β-disubstituted nitroolefins **116** to *N*-Boc aldimines **43b** (aka aza-Henry reaction) (Scheme 2.60). First, this reaction is remarkable for several reasons: (i) the deprotonation occurs at the γ position of the nitroolefins **116**, which is difficult to perform using standard bases, and (ii) the addition is totally regioselective in favor of the α position. These two facts clearly underline the unique behavior of the betaine catalysts over others catalysts. It is important to note that the substituents at the 3,3'-positions of the binaphthyl ring revealed to be crucial both in terms of yield and enantiocontrol. Indeed, while catalyst **115b** was found to be ineffective, catalyst **115a** exhibited promising yield, dia- and enantioselectivity (21%, 13:1 d.r. *anti/syn* and 77% *ee*). The best results were obtained using betaine **115c** encompassing two $3,5\text{-}(CF_3)_2C_6H_3$ units at the 3,3'-positions allowing to reach excellent level of yield, dia- and enantioselectivity (90%, >20:1 d.r. *anti/syn* and 96% *ee*). By implementing betaine **115c**, α-aryl β-methyl nitroolefins (**116a**, R = H) were added to various *N*-Boc aldimines **43b** affording the α-addition product **117** in 64–97% yields and high stereoselectivities (*anti/syn*: 5:1 to >20:1; 92–99% *ee*). Substituted aromatic ring at the α-position and longer alkyl chains at the β-position were also well tolerated provided the addition product **117** in comparable yields and stereoselectivities. In order to determine the absolute and relative configuration of the addition product, derivatization into a known α,β-diamino acid **118** was undertaken allowing to attribute the (1*S*,2*S*) configuration.

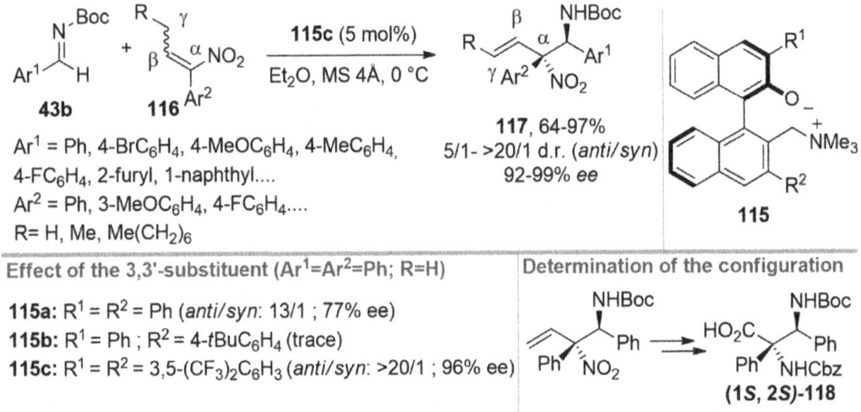

Effect of the 3,3'-substituent (Ar¹=Ar²=Ph; R=H)	Determination of the configuration

115a: R¹ = R² = Ph (*anti/syn*: 13/1 ; 77% ee)
115b: R¹ = Ph ; R² = 4-*t*BuC₆H₄ (trace)
115c: R¹ = R² = $3,5\text{-}(CF_3)_2C_6H_3$ (*anti/syn*: >20/1 ; 96% ee)

Scheme 2.60.

In 2011, Gong *et al.* reported a new class of (bis)betaine catalysts **119** made up of two *Cinchona* alkaloid ammonium linked together by a chiral binaphthoxide backbone (Scheme 2.61) [ZHA 12]. In order to evaluate the catalytic potential of these new betaines, the authors chose a Mannich reaction between azlactones **120** and aliphatic imines **43c**. The influence of the alkaloid part of the catalyst was examined with model substrates ($R^2 = n$Bu, $Ar^2 = 2$-naphthyl, $R^1 = Ph(CH_2)_2$, $Ar^1 = 4\text{-}MeC_6H_4$) revealing a high catalytic activity independently of the nature of the ammonium. Nevertheless, a significant variation both in terms of dia- and enantioselectivities was measured, the best results being obtained with the quinidinium moiety **119a** (d.r. *anti/syn*: 2.7:1, 96% *ee*). By applying this bis-quinidinium betaine, low to moderate diastereoselectivies (2:1–6:1), high yields (76–99%) and particularly excellent enantioselectivties (81–99% *ee*) were obtained for a wide range of *N*-sulfonyl aliphatic imines **43c** and 4-alkyl-2-arylazlactones **120**. To account for the excellent level of enantioinduction, the authors put forward the formation of a Brønsted acidic nucleophilic species that could activate the electrophile partner thanks to H-bonding. Lastly, note that the efficiency of the methodology was demonstrated during the course of the synthesis of a highly functionalized chiral caprolactone.

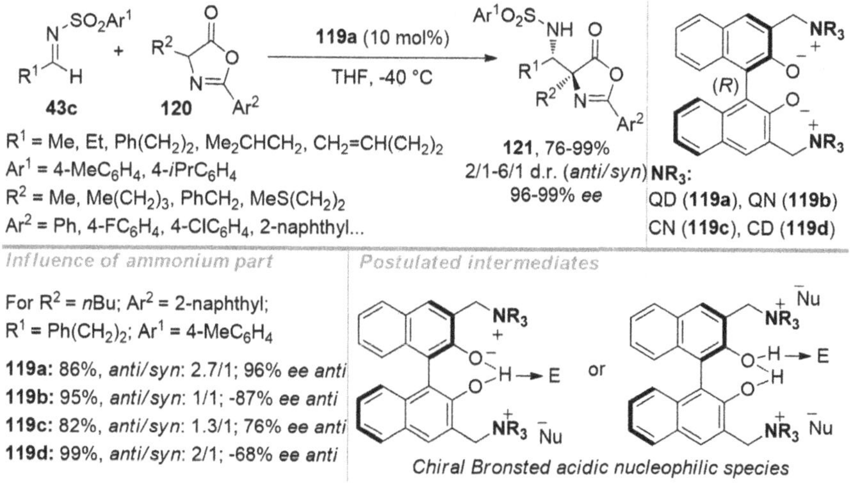

Scheme 2.61.

2.3.3. *Reactions based upon type 2 mechanism*

If one considers the nucleophilic properties of the aryloxides, type 2 mechanism based reaction should take place. A representative reaction involving such a mechanism is the Steglich rearrangement (*O* to *C* intramolecular acyl transfer). In 2010, Ooi *et al.* elegantly exploited their C_1-symmetric betaine catalysts **115** to the Steglich rearrangement of 5-oxazoly 2,2,2-trichloroethyl carbonate derivatives **122** (Scheme 2.62) [URA 10a]. The authors postulated that the observed high level of reactivity and stereocontrol could result from a pseudo-intramolecular acyl transfer occurring because of the zwitterionic nature of the betaine. The advantage of using a betaine instead of a bimolecular ammonium aryloxide system was put forward by the authors, especially from a reactivity point of view. Indeed, the betaine would render the rate-determining C–C bond forming pseudo-intramolecular reaction and would thereby provide a more reactive acylating agent than the electronically neutral aryl ester that would have been formed using an ammonium aryloxide salt. From a mechanistic point of view, after a first activation step of the substrate **122** by the betaine **115**, a new ion pair encompassing the acyl source would be formed. The ammonium part of this new ion pair was characterized by ESI-MS, thus corroborating this scenario. The C-acylation would then occur in a pseudo-intramolecular fashion (route A), thus providing the rearranged product **123** with concomitant regeneration of the betaine **115**. Nevertheless, the authors pointed out the crucial need for the reaction to be conducted at low concentration of substrate to maintain an optimal level of stereoinduction. This phenomenon was attributed to a possible competitive intermolecular pathway (route B) where the substrate **122** acts as an acyl donor instead of intermediate **124**, this process being less stereoselective than the intramolecular one. Regarding the structure of the betaine **115**, the substitution pattern of the 3,3'-positions plays a crucial role. Although catalyst **115a** ($R^1 = R^2 = Ph$) remains totally inactive, catalyst **115d** ($R^1 = H$, $R^2 = Ph$; 93%, 93% ee) proved to be quite efficient both in terms of yield and enantioselectivity, the best result being obtained with catalyst **115e** ($R^1 = H$, $R^2 = 3,5\text{-}(CF_3)_2C_6H_3$; 94%, 97% ee). Catalysts **115e** was then applied for studying the scope of this Steglich rearrangement, and excellent yields and enantioselectivties (91–99% and 94–97%, respectively) were obtained for a series of azlactone enol carbonate derivatives **122**.

Several years later, the same group reported on an innovative reaction dealing with a Steglich-type rearrangement of oxindole-derived vinylic carbonates **125** (Scheme 2.63) [URA 12a]. In this reaction, upon the addition of an additional external electrophile, such as aldehydes **39d**, transiently generated oxindole enolate forms an aldol adduct, which is eventually acylated in a pseudo-intramolecular fashion to provide the O-acylated aldol product **126**. This last acylation step, in addition to permitting the regeneration of the catalyst, also prevents the retro-aldolization reaction, which is a nice way of preventing the erosion of the enantioselectivity [OOI 04c]. Initial experiments conducted with catalysts **115d,f** using O-trichloroethoxycarbonyl enolate **125a** and benzaldehyde **39b** as model substrates provided the desired product **126a** in fair to good dia- and enantioselectivities (6:1–20:1 and 34–89% ee) while contaminated by variable amounts of undesired Steglich rearrangement product **127** (4:1–13:1). This problem could be circumvented by a modification of the carbonate substituent ($R^2 = 3,5$-$(CF_3)_2C_6H_3CH_2$) and the use of catalyst **115g**. This reaction conditions allowed not only to solve the product distribution issue (>20:1) but also to improve the enantioselectivity (95% ee) when maintaining the same level of diastereoselectivity (>20:1). These optimal conditions were then successfully applied to other oxindole-derived vinylic carbonates **125** and various aromatic aldehydes **39d** providing the corresponding O-acetylated aldol products **126** in excellent yields (73–96%) and stereoselectivities (dr 10:1–20:1, 94–97% ee). The absolute configuration of the addition product was attributed by X-ray crystallographic analysis after transformation into an indoline derivative.

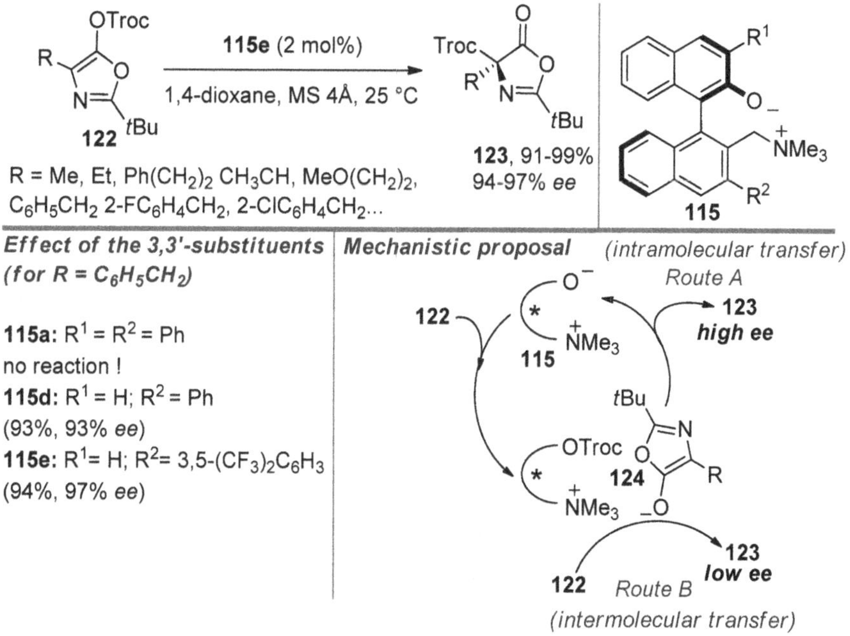

Scheme 2.62.

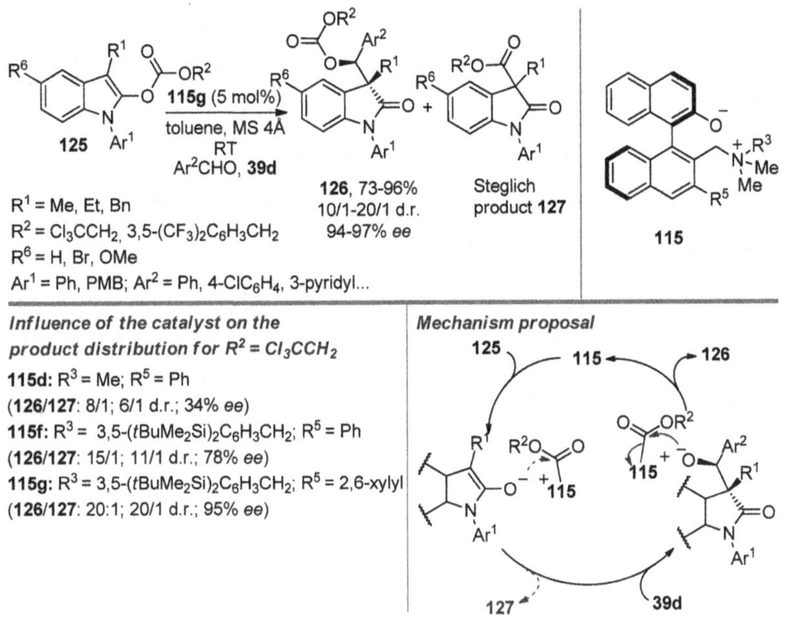
125
115g (5 mol%)
toluene, MS 4Å
RT
Ar²CHO, 39d
126, 73-96%
10/1-20/1 d.r.
94-97% ee
Steglich
product 127
115
R¹ = Me, Et, Bn
R² = Cl₃CCH₂, 3,5-(CF₃)₂C₆H₃CH₂
R⁶ = H, Br, OMe
Ar¹ = Ph, PMB; Ar² = Ph, 4-ClC₆H₄, 3-pyridyl...
Influence of the catalyst on the product distribution for R² = Cl₃CCH₂
115d: R³ = Me; R⁵ = Ph
(126/127: 8/1; 6/1 d.r.; 34% ee)
115f: R³ = 3,5-(tBuMe₂Si)₂C₆H₃CH₂; R⁵ = Ph
(126/127: 15/1; 11/1 d.r.; 78% ee)
115g: R³ = 3,5-(tBuMe₂Si)₂C₆H₃CH₂; R⁵ = 2,6-xylyl
(126/127: 20:1; 20/1 d.r.; 95% ee)
Mechanism proposal
125
115
126
127
39d

Scheme 2.63.

2.4. Conclusion

Although the chemistry of chiral quaternary ammonium salts as organocatalysts has already been studied by several exhaustive reviews, we wanted to provide the reader with basic background regarding such catalysts both from a mechanistic point of view (PTC and cooperative ion pairing strategies) and with significant and recent examples illustrating the powerfulness of these approaches. We hope that the reader will be convinced by the high efficiency generally observed for a wide variety of the reactions ranging from the classical alkylation of glycine imines to the more original and recent cycloaddition reactions. Despite the great achievements obtained in this domain in terms of enantioselective chemical transformations, there is still some more work to be done especially in terms of catalyst design. Indeed, an important number of chiral quaternary ammonium salts (*Cinchona* alkaloids derivatives, *N*-spiro ammonium binaphtyl or tartaric-derived TADDOL backbones, betaines, etc.) were synthesized and successfully evaluated in numerous asymmetric reactions. However, a fine tuning of the catalyst backbone is usually required to obtain high levels of induction for a given reaction. Moreover, such catalysts generally require several linear synthetic steps for their synthesis. This makes the design of more efficient, general and accessible chiral ammonium salts markedly desirable, which we assume to be the "holy grail" of this field of research in years to come.

2.5. Bibliography

[AHR 00] AHRENDT K.A., BORTHS C.J., MACMILLAN D.W.C., "New strategies for organic catalysis: the first highly enantioselective organocatalytic diels–alder reaction", *Journal of the American Chemical Society*, vol. 122, p. 4243, 2000.

[AIR 96] AIRES-DE-SOUSA J., LOBO A.M., PRABHAKAR S., "Characterization of methicillin-resistant Staphylococcus aureus isolates from Portuguese hospitals by multiple genotyping methods", *Tetrahedron Letters*, vol. 37, p. 3183, 1996.

[AIR 01] AIRES-DE-SOUSA J., PRABHAKAR S., LOBO A.M. *et al.*, "Asymmetric synthesis of N-aryl aziridines", *Tetrahedron: Asymmetry*, vol. 12, p. 3349, 2001.

[ALC 98] ALCARAZ L., MACDONALD G., RAGOT J.P. *et al.*, "Manumycin A: synthesis of the (+)-enantiomer and revision of stereochemical assignment", *Journal of Organic Chemistry*, vol. 63, p. 3526, 1998.

[AND 93] ANDO A., MIURA T., TATEMATSU T. *et al.*, "Chiral quarternary ammonium fluoride. A new reagent for catalytic asymmetric aldol reactions", *Tetrahedron Letters*, vol. 34, p. 1507, 1993.

[AND 08] ANDRUS M.B., YE Z., "Phase-transfer catalyzed glycolate conjugate addition", *Tetrahedron Letters*, vol. 49, p. 534, 2008.

[ARA 04] ARAI S., HASEGAWA K., NISHIDA A., "One-pot synthesis of α-diazo-β-hydroxyesters under phase-transfer catalysis and application to the catalytic asymmetric aldol reaction", *Tetrahedron Letters*, vol. 45, p. 1023, 2004.

[ARM 14] ARMSTRONG R.J., SMITH M.D., "Catalytic enantioselective synthesis of atropisomeric biaryls: a cation-directed nucleophilic aromatic substitution reaction", *Angewandte Chemie International Edition*, vol. 53, p. 12822, 2014.

[BAN 08] BANDINI M., EICHHOLZER A., TRAGNI M. *et al.*, "Enantioselective phase-transfer-catalyzed intramolecular aza-Michael reaction: effective route to pyrazino-indole compounds", *Angewandte Chemie International Edition*, vol. 47, p. 3238, 2008.

[BAN 10] BANDINI M., BOTTONI A., EICHHOLZER A. *et al.*, "Asymmetric Phase-Transfer-Catalyzed Intramolecular N-Alkylation of Indoles and Pyrroles: A Combined Experimental and Theoretical Investigation", *Chemistry: A European Journal*, vol. 16, p. 12462, 2010.

[BER 05] BERKESSEL A., GRÖGER H., *Asymmetric Organocatalysis: From Biomimetic Concepts to Applications in Asymmetric Synthesis*, Wiley-VCH Verlag GmbH & Co. KGaA, Weinheim, Germany, 2005.

[BER 07] BERKESSEL A., GUIXÀ M., SCHMIDT F. *et al.*, "Highly Enantioselective Epoxidation of 2-Methylnaphthoquinone (Vitamin K3) mediated by new cinchona alkaloid phase-transfer catalysts", *Chemistry: A European Journal*, vol. 13, p. 4483, 2007.

[BER 10] BERNAL P., FERNÁNDEZ R., LASSALETTA J.M., "Organocatalytic asymmetric cyanosilylation of nitroalkenes", *Chemistry: A European Journal*, vol. 16, p. 7714, 2010.

[BER 12] BERNARDI L., INDRIGO E., POLLICINO S. *et al.*, "Organocatalytic trifluoromethylation of imines using phase-transfer catalysis with phenoxides. A general platform for catalytic additions of organosilanes to imines", *Chemical Communications*, vol. 48, p. 1428, 2012.

[BHU 02] BHUNNOO R.A., HU Y., LAINE D.I. *et al.*, "An asymmetric phase-transfer dihydroxylation reaction", *Angewandte Chemie International Edition*, vol. 41, p. 3479, 2002.

[BLU 01] BLUET G., CAMPAGNE J.-M., "Catalytic Asymmetric Vinylogous Mukaiyama-Aldol (CAVM) reactions: the enolate activation", *Journal of Organic Chemistry*, vol. 66, p. 4293, 2001.

[BOE 02] BOECKMAN R.K., CLARK T.J., SHOOK B.C., "A practical enantioselective total synthesis of the bengamides B, E, and Z", *Organic Letters*, vol. 4, p. 2109, 2002.

[BRI 12] BRIÈRE J.-F., OUDEYER S., DALLA V. *et al.*, "Recent advances in cooperative ion pairing in asymmetric organocatalysis", *Chemical Society Reviews*, vol. 41, p. 1696, 2012.

[BRO 01] BROWN R.C.D., KEILY J.F., "Asymmetric permanganate-promoted oxidative cyclization of 1,5-dienes by using chiral phase-transfer catalysis", *Angewandte Chemie International Edition*, vol. 40, p. 4496, 2001.

[CAN 02] CANNIZZARO C.E., HOUK K.N., "Magnitudes and chemical consequences of R3N+−C−H···OC hydrogen bonding", *Journal of the American Chemical Society*, vol. 124, p. 7163, 2002.

[CAS 03] CASTLE S.L., SRIKANTH G.S.C., "Catalytic asymmetric synthesis of the central tryptophan residue of celogentin c", *Organic Letters*, vol. 5, p. 3611, 2003.

[CAS 09] CASSANI C., BERNARDI L., FINI F. *et al.,* "Catalytic asymmetric mannich reactions of sulfonylacetates", *Angewandte Chemie International Edition*, vol. 48, p. 5694, 2009.

[CHI 07] CHIDAMBARAM M., SONAVANE S.U., DE LA ZERDA J. *et al.*, "Didecyldimethylammonium bromide (DDAB): a universal, robust, and highly potent phase-transfer catalyst for diverse organic transformations", *Tetrahedron*, vol. 63, p. 7696, 2007.

[CLA 13a] CLARAZ A., LANDELLE G., OUDEYER S. *et al.*, "Asymmetric organocatalytic protonation of silyl enolates catalyzed by simple and original betaines derived from cinchona alkaloids", *European Journal of Organic Chemistry*, p. 7693, 2013.

[CLA 13b] CLARAZ A., OUDEYER S., LEVACHER V., "Chiral quaternary ammonium Aryloxide/N,O-Bis(trimethyl- silyl)acetamide combination as efficient organocatalytic system for the direct vinylogous aldol reaction of (5H)-Furan-2-one derivatives", *Advanced Synthesis & Catalysis*, vol. 355, p. 841, 2013.

[CLA 13c] CLARAZ A., OUDEYER S., LEVACHER V., "Enantioselective desymmetrization of prochiral ketones via an organocatalytic deprotonation process", *Tetrahedron: Asymmetry*, vol. 24, p. 764, 2013.

[COL 75] COLONNA S., FORNASIER R., "N-Dodecyl-N-methylephedrinium Bromide: a specific catalyst for the borohydride reduction of carbonyl compounds under phase-transfer conditions", *Synthesis*, vol. 1975, p. 531, 1975.

[COL 78] COLONNA S., HIEMSTRA H., WYNBERG H., "Asymmetric induction in the base-catalysed Michael addition of nitromethane to chalcone" *Journal of Chemical Society, Chemical Communications*, p. 238, 1978.

[COL 81] COLONNA S., RE A., WYNBERG H., "Asymmetric induction in the Michael reaction by means of chiral phase-transfer catalysts derived from cinchona and ephedra alkaloids", *Journal of the Chemical Society*, Perkin Transactions 1, p. 547, 1981.

[COR 97] COREY E.J., XU F., NOE M.C., "A rational approach to catalytic enantioselective enolate alkylation using a structurally rigidified and defined chiral quaternary ammonium salt under phase transfer conditions", *Journal of the American Chemical Society*, vol. 119, p. 12414, 1997.

[COR 99] COREY E.J., ZHANG F.Y., "Mechanism and conditions for highly enantioselective epoxidation of alpha,beta-enones using charge-accelerated catalysis by a rigid quaternary ammonium salt", *Organic Letters*, vol. 1, p. 1287, 1999.

[DAL 13] DALKO P.I., *Comprehensive Enantioselective Organocatalysis*, Wiley-VCH Verlag GmbH & Co. KGaA, Weinheim, 2013.

[DE 13] DE FREITAS MARTINS E., PLIEGO J.R., "Unraveling the Mechanism of the Cinchoninium Ion Asymmetric Phase-Transfer-Catalyzed Alkylation Reaction", *ACS Catalysis*, vol. 3, p. 613, 2013.

[DEN 11a] DENMARK S.E., GOULD N.D., WOLF L.M., "A Systematic Investigation of Quaternary Ammonium Ions as Asymmetric Phase-Transfer Catalysts. Application of Quantitative Structure Activity/Selectivity Relationships", *Journal of Organic Chemistry*, vol. 76, p. 4337, 2011.

[DEN 11b] DENMARK S.E., GOULD N.D., WOLF L.M., "A systematic investigation of quaternary ammonium ions as asymmetric phase-transfer catalysts. Synthesis of catalyst libraries and evaluation of catalyst activity", *Journal of Organic Chemistry*, vol. 76, p. 4260, 2011.

[DOL 84] DOLLING U.H., DAVIS P., GRABOWSKI E.J.J., "Efficient catalytic asymmetric alkylations. 1. Enantioselective synthesis of (+)-indacrinone via chiral phase-transfer catalysis", *Journal of the American Chemical Society*, vol. 106, p. 446, 1984.

[DUA 15] DUAN S., LI S., YE X. *et al.*, "Enantioselective Synthesis of Dialkylated α-Hydroxy carboxylic acids through asymmetric phase-transfer catalysis", *Journal of Organic Chemistry*, vol. 80, p. 7770, 2015.

[FIA 75] FIAUD J.C., "Asymmetric alkylation reaction using a chiral phase transfer catalyst", *Tetrahedron Letters*, vol. 16, p. 3495, 1975.

[FIN 05] FINI F., SGARZANI V., PETTERSEN D. *et al.*, "Phase-transfer-catalyzed asymmetric Aza-Henry reaction using N-Carbamoyl Imines Generated In Situ from α-Amido Sulfones", *Angewandte Chemie International Edition*, vol. 44, p. 7975, 2005.

[FIN 08] FINI F., MICHELETTI G., BERNARDI L. *et al.*, "An easy entry to optically active [small alpha]-amino phosphonic acid derivatives using phase-transfer catalysis (PTC)", *Chemical Communications*, p. 4345, 2008.

[FIO 04] FIORAVANTI R., MASCIA M.G., PELLACANI L. *et al.*, "Cinchona alkaloids in the asymmetric synthesis of 2-(phenylsulfanyl)aziridines", *Tetrahedron*, vol. 60, p. 8073, 2004.

[FUR 08] FURUKAWA T., SHIBATA N., MIZUTA S. *et al.*, "Catalytic Enantioselective Michael Addition of 1-Fluorobis(phenylsulfonyl)methane to α,β-Unsaturated Ketones Catalyzed by Cinchona Alkaloids", *Angewandte Chemie International Edition*, vol. 47, p. 8051, 2008.

[GAS 91] GASPARSKI C.M., MILLER M.J., "Synthesis of β-hydroxy-α-amino acids by aldol condensation using a chiral phase transfer catalyst", *Tetrahedron*, vol. 47, p. 5367, 1991.

[GEL 11] GELAT F., JAYASHANKARAN J., LOHIER J.-F. *et al.*, "Organocatalytic asymmetric synthesis of sulfoxides from sulfenic acid anions mediated by a cinchona-derived phase-transfer reagent", *Organic Letters*, vol. 13, p. 3170, 2011.

[GEL 13] GELAT F., GAUMONT A.-C., PERRIO S., "Chiral non-racemic sulfoxides by asymmetric alkylation of alkanesulfenates in the presence of a chiral ammonium phase-transfer catalyst derived from Cinchona alkaloid", *Journal of Sulfur Chemistry*, vol. 34, p. 596, 2013.

[GIO 09] GIOIA C., FINI F., MAZZANTI A. *et al.*, "Organocatalytic Asymmetric Formal [3 + 2] Cycloaddition with in Situ-Generated N-Carbamoyl Nitrones", *Journal of the American Chemical Society*, vol. 131, p. 9614, 2009.

[GOD 15] GODEMERT J., OUDEYER S., LEVACHER V., "Chiral Ammonium Aryloxides: Efficient Multipurpose Basic Organocatalysts", *ChemCatChem*, vol. 8, p. 74, 2015.

[GOM 08] GOMEZ-BENGOA E., LINDEN A., LÓPEZ R. *et al.*, "Asymmetric Aza-Henry Reaction Under Phase Transfer Catalysis: An Experimental and Theoretical Study", *Journal of the American Chemical Society*, vol. 130, p. 7955, 2008.

[GUO 15] GUO J., YU S., "Enantioselective synthesis of benzoindolizidine derivatives using chiral phase-transfer catalytic intramolecular domino aza-Michael addition/alkylation", *Organic & Biomolecular Chemistry*, vol. 13, p. 1179, 2015.

[GUR 13] GURURAJA G., HERCHL R., PICHLER A. *et al.*, "Application Scope and Limitations of TADDOL-Derived Chiral Ammonium Salt Phase-Transfer Catalysts", *Molecules*, vol. 18, p. 4357, 2013.

[HAS 07] HASHIMOTO T., MARUOKA K., "Recent Development and Application of Chiral Phase-Transfer Catalysts", *Chemical Reviews*, vol. 107, p. 5656, 2007.

[HAS 10] HASHIMOTO T., FUKUMOTO K., ABE N. *et al.*, "Development of 5-silylethynyl-1,3-dioxolan-4-one as a new prochiral template for asymmetric phase-transfer catalysis", *ChemicalCommunications*, vol. 46, p. 7593, 2010.

[HE 09] HE R., SHIRAKAWA S., MARUOKA K., "Enantioselective Base-Free Phase-Transfer Reaction in Water-Rich Solvent", *Journal of the American Chemical Society*, vol. 131, p. 16620, 2009.

[HEL 76] HELDER R., HUMMELEN J.C., LAANE R.W.P.M. *et al.*, "Catalytic asymmetric induction in oxidation reactions. The synthesis of optically active epoxides", *Tetrahedron Letters*, vol. 17, p. 1831, 1976.

[HER 14] HERCHL R., WASER M., "Stereoselective cyclization reactions under phase-transfer catalysis", *Tetrahedron*, vol. 70, p. 1935, 2014.

[HIN 12] HINTERMANN L., DITTMER C., "Asymmetric Ion-Pairing Catalysis of the Reversible Cyclization of 2′-Hydroxychalcone to Flavanone: Asymmetric Catalysis of an Equilibrating Reaction", *European Journal of Organic Chemistry*, vol. 2012, p. 5573, 2012.

[HIY 75] HIYAMA T., SAWADA H., TSUKANAKA M. *et al.*, "β-hydroxyethyltrialkylammonium ion as a selective phase-transfer catalyst for dihalocyclopropanation", *Tetrahedron Letters*, vol. 16, p. 3013, 1975.

[HOR 99] HORIKAWA M., BUSCH-PETERSEN J., COREY E.J., "Enantioselective synthesis of β-hydroxy-α-amino acid esters by aldol coupling using a chiral quaternary ammonium salt as catalyst", *Tetrahedron Letters*, vol. 40, p. 3843, 1999.

[HU 09] HU X., WANG J., LI W. *et al.*, "Cinchona alkaloid-derived quaternary ammonium salt combined with NaH: a facile catalyst system for the asymmetric trifluoromethylation of ketones", *Tetrahedron Letters*, vol. 50, p. 4378, 2009.

[HUA 10] HUA M.-Q., CUI H.-F., WANG L. *et al.*, "Reversal of Enantioselectivity by Tuning the Conformational Flexibility of Phase-Transfer Catalysts", *Angewandte Chemie International Edition*, vol. 49, p. 2772, 2010.

[HUA 11] HUA M.-Q., WANG L., CUI H.-F. *et al.*, "A powerful synergistic effect for highly efficient diastereo- and enantioselective phase-transfer catalyzed conjugate additions", *Chemical Communications*, vol. 47, p. 1631, 2011.

[HUM 78] HUMMELEN J.C., WYNBERG H., "Alkaloid assisted asymmetric synthesis. IV. Additional routes to chiral epoxides", *Tetrahedron Letters*, p. 1089, 1978.

[IKU 08] IKUNAKA M., "PTC in OPRD: An Illustrative Overview", *Organic Process Research & Development*, vol. 12, p. 698, 2008.

[ISE 94] ISEKI K., NAGAI T., KOBAYASHI Y., "Asymmetric trifluoromethylation of aldehydes and ketones with trifluoromethyltrimethylsilane catalyzed by chiral quaternary ammonium fluorides", *Tetrahedron Letters*, vol. 35, p. 3137, 1994.

[IUL 10] IULIANO A., "Asymmetric activation of tropos species in the achievement of chiral inducers for enantioselective catalysis", *Tetrahedron: Asymmetry*, vol. 21, p. 1943, 2010.

[JEW 03] JEW S.-S., JEONG B.-S., LEE J.-H. *et al.*, "Highly Enantioselective Synthesis of α-Alkyl-alanines via the Catalytic Phase-Transfer Alkylation of 2-Naphthyl Aldimine tert-Butyl Ester by Using O(9)-Allyl-N(1)-2',3',4'-trifluorobenzylhydrocinchonidinium Bromide", *Journal of Organic Chemistry*, vol. 68, p. 4514, 2003.

[JEW 04] JEW S.-S., LEE Y.-J., LEE J. *et al.*, "Highly Enantioselective Phase-Transfer-Catalytic Alkylation of 2-Phenyl-2-oxazoline-4-carboxylic Acid tert-Butyl Ester for the Asymmetric Synthesis of α-Alkyl Serines", *Angewandte Chemie International Edition*, vol. 43, p. 2382, 2004.

[JEW 09] JEW S.-S., PARK H.-G., "Cinchona-based phase-transfer catalysts for asymmetric synthesis", *Chemical Communications*, p. 7090, 2009.

[JOH 12] JOHNSON K.M., RATTLEY M.S., SLADOJEVICH F. *et al.*, "A new family of cinchona-derived bifunctional asymmetric phase-transfer catalysts: application to the enantio- and diastereoselective Nitro-Mannich reaction of amidosulfones", *Organic Letters*, vol. 14, p. 2492, 2012.

[JOH 15] JOHNSTON C.P., KOTHARI A., SERGEIEVA T. *et al.*, "Catalytic enantioselective synthesis of indanes by a cation-directed 5-endo-trig cyclization", *Nature Chemistry*, vol. 7, p. 171, 2015.

[JUL 80] JULIÁ S., GINEBREDA A., GUIXER J. *et al.*, "Phase transfer catalysis using chiral catalysts. V. Asymmetric nucleophilic substitutions with C, O, N and S-anions", *Tetrahedron Letters*, vol. 21, p. 3709, 1980.

[KAG 88] KAGAN H.B., FIAUD J.C., "Kinetic Resolution", in ELIEL E.L., WILEN S.H. (eds), *Topics in Stereochemistry*, John Wiley & Sons, Hoboken, 1988.

[KAN 16] KANEKO S., KUMATABARA Y., SHIRAKAWA S., "A new generation of chiral phase-transfer catalysts", *Organic & Biomolecular Chemistry*, vol. 14, p. 5367–5376, 2016.

[KAW 09] KAWAI H., KUSUDA A., NAKAMURA S. *et al.*, "Catalytic Enantioselective Trifluoromethylation of Azomethine Imines with Trimethyl(trifluoromethyl) silane", *Angewandte Chemie International Edition*, vol. 48, p. 6324, 2009.

[KAW 13] KAWAI H., OKUSU S., YUAN Z., *et al.*, "Enantioselective Synthesis of Epoxides Having a Tetrasubstituted Trifluoromethylated Carbon Center: Methylhydrazine-Induced Aerobic Epoxidation of β,β-Disubstituted Enones", *Angewandte Chemie International Edition*, vol. 52, p. 2221, 2013.

[KIM 02] KIM D.Y., PARK E.J., "Catalytic Enantioselective Fluorination of β-Keto Esters by Phase-Transfer Catalysis Using Chiral Quaternary Ammonium Salts", *Organic Letters*, vol. 4, p. 545, 2002.

[KIM 06] KIM T.-S., LEE Y.-J., JEONG B.-S. *et al.*, "Enantioselective Synthesis of (R)- and (S)-α-Alkylcysteines via Phase-Transfer Catalytic Alkylation", *Journal of Organic Chemistry*, vol. 71, p. 8276, 2006.

[KIM 09] KIM T.-S., LEE Y.-J., LEE K. *et al.*, "Enantioselective Synthesis of (R)-α-Alkylhomoserines and (R)-α-Alkylhomocysteines via Phase-Transfer Catalytic Alkylation", *Synlett*, vol. 2009, p. 671, 2009.

[KIT 08] KITAMURA M., SHIRAKAWA S., ARIMURA Y. *et al.*, "Combinatorial Design of Simplified High-Performance Chiral Phase-Transfer Catalysts for Practical Asymmetric Synthesis of α-Alkyl- and α,α-Dialkyl-α-Amino Acids", *Chemistry: An Asian Journal*, vol. 3, p. 1702, 2008.

[LEB 03] LEBEL H., MORIN S., PAQUET V., "Alkylation of Phosphine Boranes by Phase-Transfer Catalysis", *Organic Letters*, vol. 5, p. 2347, 2003.

[LEB 15] LEBRUN S., SALLIO R., DUBOIS M. *et al.*, "Chiral Phase-Transfer-Catalyzed Intramolecular aza-Michael Reactions for the Asymmetric Synthesis of Isoindolinones", *European Journal of Organic Chemistry*, vol. 2015, p. 1995, 2015.

[LEE 05] LEE Y.-J., LEE J., KIM M.-J. *et al.*, "Highly Enantioselective Synthesis of (R)-α-Alkylserines via Phase-Transfer Catalytic Alkylation of o-Biphenyl-2-oxazoline-4-carboxylic Acid tert-Butyl Ester Using Cinchona-Derived Catalysts", *Organic Letters*, vol. 7, p. 1557, 2005.

[LEE 15a] LEE H.-J., CHO C.-W., "Enantioselective Phase-Transfer-Catalyzed Synthesis of Chiral N-Substituted 3,3-Dinitroazetidines by Aza-Michael Reaction", *Journal of Organic Chemistry*, vol. 80, p. 11435, 2015.

[LEE 15b] LEE S.-J., BAE J.-Y., CHO C.-W., "Phase-Transfer-Catalyzed Asymmetric Synthesis of Chiral N-Substituted Pyr-azoles by Aza-Michael Reaction", *European Journal of Organic Chemistry*, vol. 2015, p. 6495, 2015.

[LI 13] LI M., WOODS P.A., SMITH M.D., "Cation-directed enantioselective synthesis of quaternary-substituted indolenines", *Chemical Science*, vol. 4, p. 2907, 2013.

[LIA 12] LIAN M., LI Z., CAI Y., "Enantioselective Photooxygenation of β-Keto Esters by Chiral Phase-Transfer Catalysis using Molecular Oxygen", *Chemistry: An Asian Journal*, vol. 7, p. 2019, 2012.

[LIU 11] LIU Y., PROVENCHER B.A., BARTELSON K.J. *et al.*, "Highly enantioselective asymmetric Darzens reactions with a phase transfer catalyst", *Chemical Science*, vol. 2, p. 1301, 2011.

[LUO 07] LUO Z.-B., HOU X.-L., DAI L.-X., "Enantioselective desymmetrization of meso-N-sulfonylaziridines with thiols", *Tetrahedron: Asymmetry*, vol. 18, p. 443, 2007.

[LUO 13] LUO J., WU W., XU L.-W. *et al.*, "Enantioselective direct fluorination and chlorination of cyclic β-ketoesters mediated by phase-transfer catalysts", *Tetrahedron Letters*, vol. 54, p. 2623, 2013.

[LYG 98] LYGO B., WAINWRIGHT P.G., "Asymmetric phase-transfer mediated epoxidation of α,β-unsaturated ketones using catalysts derived from Cinchona alkaloids", *Tetrahedron Letters*, vol. 39, p. 1599, 1998.

[LYG 99a] LYGO B., CROSBY J., PETERSON J.A., "Enantioselective alkylation of alanine-derived imines using quaternary ammonium catalysts", *Tetrahedron Letters*, vol. 40, p. 8671, 1999.

[LYG 99b] LYGO B., WAINWRIGHT P.G., "Phase-transfer catalyzed asymmetric epoxidation of enones using N-anthracenylmethyl-substituted cinchona alkaloids", *Tetrahedron*, vol. 55, p. 6289, 1999.

[LYG 02] LYGO B., DANIEL C.M., "Asymmetric epoxidation via phase-transfer catalysis: direct conversion of allylic alcohols into α,β-epoxyketones", *Chemical Communications*, p. 2360, 2002.

[LYG 03] LYGO B., ALLBUTT B., JAMES S.R., "Identification of a highly effective asymmetric phase-transfer catalyst derived from α-methylnaphthylamine", *Tetrahedron Letters*, vol. 44, p. 5629, 2003.

[LYG 04a] LYGO B., ALLBUTT B., "Asymmetric PTC Alkylation of Glycine Imines: Variation of the Imine Ester Moiety", *Synlett*, vol. 2004, p. 0326, 2004.

[LYG 04b] LYGO B., ANDREWS B.I., "Asymmetric Phase-Transfer Catalysis Utilizing Chiral Quaternary Ammonium Salts: Asymmetric Alkylation of Glycine Imines", *Accounts of Chemical Research*, vol. 37, p. 518, 2004.

[LYG 05] LYGO B., ALLBUTT B., KIRTON E.H.M., "Asymmetric Michael addition of glycine imines via quaternary ammonium ion catalysis", *Tetrahedron Letters*, vol. 46, p. 4461, 2005.

[LYG 09a] LYGO B., ALLBUTT B., BEAUMONT D.J. *et al.*, "Chiral Dibenzazepinium Halide Phase-BarrTransfyer Catalysts Lygo,*", *Synlett*, p. 675, 2009.

[LYG 09b] LYGO B., BEYNON C., LUMLEY C. *et al.*, "Co-catalyst enhancement of enantioselective PTC Michael additions involving glycine imines", *Tetrahedron Letters*, vol. 50, p. 3363, 2009.

[LYK 13] LYKKE L., HALSKOV K.S., CARLSEN B.D. *et al.*, "Catalytic Asymmetric Diaziridination", *Journal of the American Chemical Society*, vol. 135, p. 4692, 2013.

[MA 07] MA B., PARKINSON J.L., CASTLE S.L., "Novel Cinchona alkaloid derived ammonium salts as catalysts for the asymmetric synthesis of β-hydroxy α-amino acids via aldol reactions", *Tetrahedron Letters*, vol. 48, p. 2083, 2007.

[MA 11] MA T., FU X., KEE C.W. *et al.*, "Pentanidium-Catalyzed Enantioselective Phase-Transfer Conjugate Addition Reactions", *Journal of the American Chemical Society*, vol. 133, p. 2828, 2011.

[MAC 98] MACDONALD G., ALCARAZ L., LEWIS N.J. *et al.*, "Asymmetric synthesis of the mC7N core of the manumycin family: preparation of (+)-MT 35214 and a formal total synthesis of (-)-alisamycin", *Tetrahedron Letters*, vol. 39, p. 5433, 1998.

[MAC 09] MACIVER E.E., THOMPSON S., SMITH M.D., "Catalytic Asymmetric 6π Electrocyclization: Enantioselective Synthesis of Functionalized Indolines", *Angewandte Chemie International Edition*, vol. 48, p. 9979, 2009.

[MAC 08] MACMILLAN D.W.C., "The advent and development of organocatalysis", *Nature*, vol. 455, p. 304, 2008.

[MAH 10] MAHÉ O., DEZ I., LEVACHER V. *et al.*, "Enantioselective phase-transfer catalysis: synthesis of pyrazolines", *Angewandte Chemie International Edition*, vol. 39, p. 7072, 2010.

[MAH 12] MAHÉ O., DEZ I., LEVACHER V. *et al.*, "Enantioselective synthesis of bio-relevant 3,5-diaryl pyrazolines", *Organic & Biomolecular Chemistry*, vol. 10, p. 3946, 2012.

[MAI 11] MAITY P., LEPORE S.D., "Catalytic synthesis of nonracemic azaproline derivatives by cyclization of β-alkynyl hydrazines under kinetic resolution conditions", *Angewandte Chemie International Edition*, vol. 50, p. 8338, 2011.

[MON 11] MONGE D., JENSEN K.L., MARÍN I. *et al.*, "Synthesis of 1,2,4-Triazolines: Base-Catalyzed Hydrazination/Cyclization Cascade of α-Isocyano Esters and Amides", *Organic Letters*, vol. 13, p. 328, 2011.

[MAR 10a] MARCELLI T., HIEMSTRA H., "Cinchona Alkaloids in Asymmetric Organocatalysis", *Synthesis*, vol. 2010, p. 1229, 2010.

[MAR 07a] MARIANACCI O., MICHELETTI G., BERNARDI L. *et al.*, "Organocatalytic Asymmetric Mannich Reactions with N-Boc and N-Cbz Protected α-Amido Sulfones (Boc: tert-Butoxycarbonyl, Cbz: Benzyloxycarbonyl)", *Chemistry: A European Journal*, vol. 13, p. 8338, 2007.

[MAR 07b] MARUOKA K., OOI T., KANO T., "Design of chiral organocatalysts for practical asymmetric synthesis of amino acid derivatives", *Chemical Communications*, p. 1487, 2007.

[MAR 08a] MARUOKA K., "Practical Aspects of Recent Asymmetric Phase-Transfer Catalysis" *Organic Process Research & Development*, vol. 12, p. 679, 2008.

[MAR 08b] MARUOKA K. (ed), *Asymmetric Phase Transfer Catalysis*, Wiley-VCH Verlag GmbH & Co. KGaA, Weinheim, Germany, 2008.

[MAR 10b] MARUOKA K., "Highly practical amino acid and alkaloid synthesis using designer chiral phase transfer catalysts as high-performance organocatalysts", *The Chemical Record*, vol. 10, p. 254, 2010.

[MAS 88] MASUI M., ANDO A., SHIOIRI T., "New method and reagents in organic synthesis. 75. Asymmetric synthesis of α-hydroxy ketones using chiral phase transfer catalysts", *Tetrahedron Letters*, vol. 29, p. 2835, 1988.

[MAT 10] MATOBA K., KAWAI H., FURUKAWA T. *et al.*, "Enantioselective synthesis of trifluoromethyl-substituted 2-isoxazolines: asymmetric hydroxylamine/enone cascade reaction", *Angewandte Chemie International Edition*, vol. 49, p. 5762, 2010.

[MIN 08] MINAKATA S., MURAKAMI Y., TSURUOKA R. *et al.*, "Catalytic aziridination of electron-deficient olefins with an N-chloro-N-sodio carbamate and application of this novel method to asymmetric synthesis", *Chemical Communications*, p. 6363, 2008.

[MIZ 07] MIZUTA S., SHIBATA N., AKITI S., *et al.*, "Cinchona Alkaloids/TMAF Combination-Catalyzed Nucleophilic Enantioselective Trifluoromethylation of Aryl Ketones", *Organic Letters*, vol. 9, p. 3707, 2007.

[MOM 09] MOMO R.D., FINI F., BERNARDI L. *et al.*, "Asymmetric Synthesis of α,β-Diaminophosphonic Acid Derivatives with a Catalytic Enantioselective Mannich Reaction", *Advanced Synthesis & Catalysis*, vol. 351, p. 2283, 2009.

[MUK 06] MUKAIYAMA T., NAGAO H., YAMANE Y., "Diastereo- and Enantioselective Tandem Michael Addition and Lactonization Catalyzed by Chiral Quaternary Ammonium Phenoxide: Stereoselective Synthesis of the Two Enantiomers by Using a Single Chiral Source", *Chemical Letters*, vol. 35, p. 916, 2006.

[MUK 07] MUKAIYAMA T., NAGAO H., YAMANE Y., "Efficient synthesis of Substituted 3-Amino-3,4-dihydropyran-2-ones Diastereo and Enantioselective Tandem Michael addition and lactonization between α,β-Unsaturated Ketones and Glycine-Derived Silyl Enolates using a Chiral Quaternary Ammonium Phenoxide", *Heterocycles*, vol. 72, p. 553, 2007.

[MUR 05] MURUGAN E., SIVA A., "Synthesis of Asymmetric N-Arylaziridine Derivatives Using a New Chiral Phase-Transfer Catalyst", *Synthesis*, vol. 12, p. 2022, 2005.

[MUR 11] MURAKAMI Y., TAKEDA Y., MINAKATA S., "Diastereoselective Aziridination of Chiral Electron-Deficient Olefins with N-Chloro-N-sodiocarbamates Catalyzed by Chiral Quaternary Ammonium Salts", *Journal of Organic Chemistry*, vol. 76, p. 6277, 2011.

[NAG 06] NAGAO H., YAMANE Y., MUKAIYAMA T., "Diastereo- and Enantioselective Synthesis of 3-Amino-3,4-dihydropyran-2-ones by Tandem Michael Addition and Lactonization Using a Chiral Quaternary Ammonium Phenoxide", *Chemical Letters*, vol. 35, p. 1398, 2006.

[NAG 07a] NAGAO H., YAMANE Y., MUKAIYAMA T., "Effective Synthesis of 5-Substituted Butenolide Derivatives by Using Cinchonidine-derived Quaternary Ammonium Phenoxide Catalyst", *Chemical Letters*, vol. 36, p. 8, 2007.

[NAG 07b] NAGAO H., YAMANE Y., MUKAIYAMA T., "Asymmetric Trifluoromethylation of Ketones with (Trifluoromethyl)trimethylsilane Catalyzed by Chiral Quaternary Ammonium Phenoxides", *Chemical Letters*, vol. 36, p. 666, 2007.

[NEL 99] NELSON A., "Asymmetric Phase-Transfer Catalysis", *Angewandte Chemie International Edition*, vol. 38, p. 1583, 1999.

[NIB 09] NIBBS A.E., BAIZE A.-L., HERTER R.M. et al., "Catalytic Asymmetric Alkylation of Substituted Isoflavanones", *Organic Letters*, vol. 11, p. 4010, 2009.

[NOV 13] NOVACEK J., WASER M., "Bifunctional Chiral Quaternary Ammonium Salt Catalysts: A Rapidly Emerging Class of Powerful Asymmetric Catalysts", *European Journal of Organic Chemistry*, vol. 2013, p. 637, 2013.

[NOV 14] NOVACEK J., WASER M., "Syntheses and Applications of (Thio)Urea-Containing Chiral Quaternary Ammonium Salt Catalysts", *European Journal of Organic Chemistry*, vol. 2014, p. 802, 2014.

[ODO 89] O'DONNELL M.J., BENNETT W.D., WU S., "The stereoselective synthesis of α-amino acids by phase-transfer catalysis", *Journal of the American Chemical Society*, vol. 111, p. 2353, 1989.

[ODO 92] O'DONNELL M.J., WU S., "A catalytic enantioselective synthesis of α-methyl amino acid derivatives by phase-transfer catalysis", *Tetrahedron: Asymmetry*, vol. 3, p. 591, 1992.

[ODO 94] O'DONNELL M.J., WU S., HUFFMAN J.C., "A new active catalyst species for enantioselective alkylation by phase-transfer catalysis", *Tetrahedron*, vol. 50, p. 4507, 1994.

[ODO 01] O'DONNELL M.J., DELGADO F., DOMÍNGUEZ E. et al., "Enantioselective solution- and solid-phase synthesis of glutamic acid derivatives via Michael addition reactions", *Tetrahedron: Asymmetry*, vol. 12, p. 821, 2001.

[ODO 04] O'DONNELL M.J., "The Enantioselective Synthesis of α-Amino Acids by Phase-Transfer Catalysis with Achiral Schiff Base Esters", *Accounts of Chemical Research*, vol. 37, p. 506, 2004.

[OHM 11] OHMATSU K., KIYOKAWA M., OOI T., "Chiral 1,2,3-Triazoliums as New Cationic Organic Catalysts with Anion-Recognition Ability: Application to Asymmetric Alkylation of Oxindoles", *Journal of the American Chemical Society*, vol. 133, p. 1307, 2011.

[OHM 12] OHMATSU K., GOTO A., OOI T., "Catalytic asymmetric Mannich-type reactions of α-cyano α-sulfonyl carbanions", *Chemical Communications*, vol. 48, p. 7913, 2012.

[OHS 04] OHSHIMA T., SHIBUGUCHI T., FUKUTA Y., *et al.*, "Catalytic asymmetric phase-transfer reactions using tartrate-derived asymmetric two-center organocatalysts", *Tetrahedron*, vol. 60, p. 7743, 2004.

[OKA 05] OKADA A., SHIBUGUCHI T., OHSHIMA T. *et al.*, "Enantio- and Diastereoselective Catalytic Mannich-Type Reaction of a Glycine Schiff Base Using a Chiral Two-Center Phase-Transfer Catalyst", *Angewandte Chemie International Edition*, vol. 44, p. 4564, 2005.

[OOI 99] OOI T., KAMEDA M., MARUOKA K., "Molecular Design of a C2-Symmetric Chiral Phase-Transfer Catalyst for Practical Asymmetric Synthesis of α-Amino Acids", *Journal of the American Chemical Society*, vol. 121, p. 6519, 1999.

[OOI 00] OOI T., TAKEUCHI M., KAMEDA M. *et al.*, "Practical Catalytic Enantioselective Synthesis of α,α-Dialkyl-α-amino Acids by Chiral Phase-Transfer Catalysis", *Journal of the American Chemical Society*, vol. 122, p. 5228, 2000.

[OOI 02a] OOI T., TAKAHASHI M., DODA K. *et al.*, "Asymmetric Induction in the Neber Rearrangement of Simple Ketoxime Sulfonates under Phase-Transfer Conditions: Experimental Evidence for the Participation of an Anionic Pathway", *Journal of the American Chemical Society*, vol. 124, p. 7640, 2002.

[OOI 02b] OOI T., TANIGUCHI M., KAMEDA M. *et al.*, "Direct Asymmetric Aldol Reactions of Glycine Schiff Base with Aldehydes Catalyzed by Chiral Quaternary Ammonium Salts", *Angewandte Chemie International Edition*, vol. 41, p. 4542, 2002.

[OOI 02c] OOI T., UEMATSU Y., KAMEDA M. *et al.*, "Conformationally Flexible, Chiral Quaternary Ammonium Bromides for Asymmetric Phase-Transfer Catalysis", *Angewandte Chemie International Edition*, vol. 41, p. 1551, 2002.

[OOI 03] OOI T., DODA K., MARUOKA K., "Designer Chiral Quaternary Ammonium Bifluorides as an Efficient Catalyst for Asymmetric Nitroaldol Reaction of Silyl Nitronates with Aromatic Aldehydes", *Journal of the American Chemical Society*, vol. 125, p. 2054, 2003.

[OOI 04a] OOI T., MARUOKA K., "Asymmetric Organocatalysis of Structurally Well-Defined Chiral Quaternary Ammonium Fluorides", *Accounts of Chemical Research*, vol. 37, p. 526, 2004.

[OOI 04b] OOI T., KAMEDA M., FUJII J.-I. *et al.*, "Catalytic Asymmetric Synthesis of a Nitrogen Analogue of Dialkyl Tartrate by Direct Mannich Reaction under Phase-Transfer Conditions", *Organic Letters*, vol. 6, p. 2397, 2004.

[OOI 04c] OOI T., KAMEDA M., TANIGUCHI M., *et al.*, "Development of Highly Diastereo- and Enantioselective Direct Asymmetric Aldol Reaction of a Glycinate Schiff Base with Aldehydes Catalyzed by Chiral Quaternary Ammonium Salts", *Journal of the American Chemical Society*, vol. 126, p. 9685, 2004.

[OOI 04d] OOI T., OHARA D., TAMURA M. *et al.*, "Design of New Chiral Phase-Transfer Catalysts with Dual Functions for Highly Enantioselective Epoxidation of α,β-Unsaturated Ketones", *Journal of the American Chemical Society*, vol. 126, p. 6844, 2004.

[OOI 06a] OOI T., FUKUMOTO K., MARUOKA K., "Construction of Enantiomerically Enriched Tertiary α-Hydroxycarboxylic Acid Derivatives by Phase-Transfer-Catalyzed Asymmetric Alkylation of Diaryloxazolidin-2,4-diones", *Angewandte Chemie International Edition*, vol. 45, p. 3839, 2006.

[OOI 06b] OOI T., MIKI T., FUKUMOTO K. *et al.*, "Asymmetric Synthesis of α-Acyl-γ-butyrolactones Possessing All-Carbon Quaternary Stereocenters by Phase-Transfer-Catalyzed Alkylation", *Advanced Synthesis & Catalysis*, vol. 348, p. 1539, 2006.

[OOI 06c] OOI T., UEMATSU Y., KAMEDA M. *et al.*, "Asymmetric phase-transfer catalysis of homo- and heterochiral quaternary ammonium salts: development and application of conformationally flexible chiral phase-transfer catalysts", *Tetrahedron*, vol. 62, p. 11425, 2006.

[OOI 07a] OOI T., KATO D., INAMURA K. *et al.*, "Practical Stereoselective Synthesis of β-Branched α-Amino Acids through Efficient Kinetic Resolution in the Phase-Transfer-Catalyzed Asymmetric Alkylations", *Organic Letters*, vol. 9, p. 3945, 2007.

[OOI 07b] OOI T., MARUOKA K., "Recent Advances in Asymmetric Phase-Transfer Catalysis" *Angewandte Chemie International Edition*, vol. 46, p. 4222, 2007.

[OYA 15] OYAIZU K., URAGUCHI D., OOI T., "Vinylogy in nitronates: utilization of alpha-aryl conjugated nitroolefins as a nucleophile for a highly stereoselective aza-Henry reaction", *Chemical Communications*, vol. 51, p. 4437, 2015.

[PAR 09] PARK Y., KANG S., LEE Y.J. *et al.*, "Highly Enantioselective Synthesis of (S)-α-Alkyl-α,β-diaminopropionic Acids via Asymmetric Phase-Transfer Catalytic Alkylation of 2-Phenyl-2-imidazoline-4-carboxylic Acid tert-Butyl Esters", *Organic Letters*, vol. 11, p. 3738, 2009.

[PAR 12] PARK Y., LEE Y.J., HONG S. *et al.*, "Highly Enantioselective Total Synthesis of (+)-Isonitramine", *Organic Letters*, vol. 14, p. 852, 2012.

[PAT 07] PATTERSON D.E., XIE S., JONES L.A. *et al.*, "Synthesis of 4-Fluoro-β-(4-fluorophenyl)-l-phenylalanine by an Asymmetric Phase-Transfer Catalyzed Alkylation: Synthesis on Scale and Catalyst Stability", *Organic Process Research & Development*, vol. 11, p. 624, 2007.

[PAT 09] PATEL S.G., WISKUR S.L., "Mechanistic investigations of the Mukaiyama aldol reaction as a two part enantioselective reaction", *Tetrahedron Letters*, vol. 50, p. 1164, 2009.

[PEN 07] PENG D., ZHOU H., LIU X. *et al.*, "Enantioselective Cyanoformylation of Aldehydes Catalyzed by a Chiral Quaternary Ammonium Salt and Triethylamine", *Synlett*, p. 2448, 2007.

[PER 14] PERILLO M., DI MOLA A., FILOSA R. *et al.*, "Cascade reactions of glycine Schiff bases and chiral phase transfer catalysts in the synthesis of α-amino acids 3-substituted phthalides or isoindolinones", *RSC Advances*, vol. 4, p. 4239, 2014.

[PLU 80] PLUIM H., WYNBERG H., "Catalytic asymmetric induction in oxidation reactions. Synthesis of optically active epoxynaphthoquinones", *Journal of Organic Chemistry*, vol. 45, p. 2498, 1980.

[POU 06] POULSEN T.B., BERNARDI L., BELL M. *et al.*, "Organocatalytic Enantioselective Nucleophilic Vinylic Substitution", *Angewandte Chemie International Edition*, vol. 45, p. 6551, 2006.

[POU 07] POULSEN T.B., BERNARDI L., ALEMÁN J. *et al.*, "Organocatalytic Asymmetric Direct α-Alkynylation of Cyclic β-Ketoesters", *Journal of the American Chemical Society*, vol. 129, p. 441, 2007.

[RAB 86] RABINOVITZ M., COHEN Y., HALPERN M., "Hydroxide Ion Initiated Reactions Under Phase Transfer Catalysis Conditions: Mechanism and Implications. New Synthetic Methods (62)", *Angewandte Chemie International Edition*, vol. 25, p. 960, 1986.

[REE 93] REETZ M.T., BINGEL C., HARMS K., "Structure of carbanions having carbocations as counterions", *Journal of Chemical Society, Chemical Communications*, p. 1558, 1993.

[REE 99] REETZ M.T., HÜTTE S., GODDARD R., "Tetrabutylammonium Phenyl(phenylsulfonyl)methylide: A Chiral Metal-free 'Carbanion'", *European Journal of Organic Chemistry*, vol. 1999, p. 2475, 1999.

[ROD 10] RODRIGUES A., WLADISLAW B., VITTA C.D. *et al.*, "Asymmetric phase-transfer catalytic sulfanylation of some 2-methylsulfinyl cyclanones. Modeling of the stereochemical course of the aldol reaction of (SS,2S)-2-methylsulfinyl-2-methylsulfanylcyclohexanone", *Tetrahedron Letters*, vol. 51, p. 5344, 2010.

[SAL 13] SALLIO R., LEBRUN S., SCHIFANO-FAUX N. *et al.*, "Enantioenriched Isoindolinones from Chiral Phase-Transfer-Catalyzed Intramolecular Aza-Michael Reactions", *Synlett*, vol. 24, p. 1785, 2013.

[SAN 08] SANO D., NAGATA K., ITOH T., "Catalytic Asymmetric Hydroxylation of Oxindoles by Molecular Oxygen Using a Phase-Transfer Catalyst", *Organic Letters*, vol. 10, p. 1593, 2008.

[SEA 05] SEAYAD J., LIST B., "Asymmetric organocatalysis", *Organic & Biomolecular Chemistry*, vol. 3, p. 719, 2005.

[SHA 15] SHARMA K., WOLSTENHULME J.R., PAINTER P.P. *et al.*, "Cation-Controlled Enantioselective and Diastereoselective Synthesis of Indolines: An Autoinductive Phase-Transfer Initiated 5-endo-trig Process", *Journal of the American Chemical Society*, vol. 137, p. 13414, 2015.

[SHI 02] SHIBUGUCHI T., FUKUTA Y., AKACHI Y. *et al.*, "Development of new asymmetric two-center catalysts in phase-transfer reactions", *Tetrahedron Letters*, vol. 43, p. 9539, 2002.

[SHI 07] SHIBUGUCHI T., MIHARA H., KURAMOCHI A. *et al.*, "Catalytic Asymmetric Phase-Transfer Michael Reaction and Mannich-Type Reaction of Glycine Schiff Bases with Tartrate-Derived Diammonium Salts", *Chemistry: An Asian Journal*, vol. 2, p. 794, 2007.

[SHI 11] SHIRAKAWA S., LIU K., ITO H. *et al.*, "Phase-Transfer-Catalyzed Asymmetric Synthesis of 1,1-Disubstituted Tetrahydroisoquinolines", *Advanced Synthesis & Catalysis*, vol. 353, p. 2614, 2011.

[SHI 12] SHIRAKAWA S., LIU K., MARUOKA K., "Catalytic Asymmetric Synthesis of Axially Chiral o-Iodoanilides by Phase-Transfer Catalyzed Alkylations", *Journal of the American Chemical Society*, vol. 134, p. 916, 2012.

[SHI 13a] SHIRAKAWA S., MARUOKA K., "Recent Developments in Asymmetric Phase-Transfer Reactions", *Angewandte Chemie International Edition*, vol. 52, p. 4312, 2013.

[SHI 13b] SHIRAKAWA S., MARUOKA K., "Chiral onimum salts", in DALKO P. (ed.), *Comprehensive Enantioselective Organocatalysis*, Wiley-VCH Verlag GmbH & Co. KGaA, Weinheim, 2013.

[SHI 13c] SHIRAKAWA S., WU X., MARUOKA K., "Kinetic resolution of axially chiral 2-amino-1,1'-biaryls by phase-transfer-catalyzed N-allylation", *Angewandte Chemie International Edition*, vol. 52, p. 14200, 2013.

[SHI 14a] SHIRAKAWA S., MAKINO H., YOSHIDOME T. *et al.*, "Effect of Brønsted acid co-catalyst in asymmetric conjugate addition of 3-aryloxindoles to maleimide under base-free phase-transfer conditions", *Tetrahedron*, vol. 70, p. 7128, 2014.

[SHI 14b] SHIRAKAWA S., MARUOKA K., "Asymmetric phase-transfer reactions under base-free neutral conditions", *Tetrahedron Letters*, vol. 55, p. 3833, 2014.

[SHI 14c] SHIRAKAWA S., WANG L., HE R. *et al.*, "A Base-Free Neutral Phase-Transfer Reaction System", *Chemistry: An Asian Journal*, vol. 9, p. 1586, 2014.

[SHI 15] SHIRAKAWA S., LIU S., KANEKO S. *et al.*, "Tetraalkylammonium Salts as Hydrogen-Bonding Catalysts", *Angewandte Chemie International Edition*, vol. 54, p. 15767, 2015.

[SIM 15] SIM S.-B.D., WANG M., ZHAO Y., "Phase-Transfer-Catalyzed Enantioselective α-Hydroxylation of Acyclic and Cyclic Ketones with Oxygen", *ACS Catalysis*, vol. 5, p. 3609, 2015.

[STA 71] STARKS C.M., "Phase-transfer catalysis. I. Heterogeneous reactions involving anion transfer by quaternary ammonium and phosphonium salts", *Journal of the American Chemical Society*, vol. 93, p. 195, 1971.

[TAN 15] TAN J., YASUDA N., "Contemporary Asymmetric Phase Transfer Catalysis: Large-Scale Industrial Applications", *Organic Process Research & Development*, vol. 19, p. 1731, 2015.

[TAN 12] TANZER E.-M., SCHWEIZER W.B., EBERT M.-O. *et al.*, "Designing Fluorinated Cinchona Alkaloids for Enantioselective Catalysis: Controlling Internal Rotation by a Fluorine-Ammonium Ion gauche Effect (φNCCF)", *Chemistry: A European Journal*, vol. 18, p. 2006, 2012.

[TIF 15] TIFFNER M., NOVACEK J., BUSILLO A. *et al.*, "Design of chiral urea-quaternary ammonium salt hybrid catalysts for asymmetric reactions of glycine Schiff bases", *RSC Advances*, vol. 5, p. 78941, 2015.

[TOM 10] TOMOOKA K., UEHARA K., NISHIKAWA R. *et al.*, "Enantioselective Synthesis of Planar Chiral Organonitrogen Cycles", *Journal of the American Chemical Society*, vol. 132, p. 9232, 2010.

[TOZ 06a] TOZAWA T., YAMANE Y., MUKAIYAMA T., "Enantioselective Synthesis of 3,4-Dihydropyran-2-ones by Chiral Quaternary Ammonium Phenoxide-catalyzed Tandem Michael Addition and Lactonization", *Chemical Letters*, vol. 35, p. 56, 2006.

[TOZ 06b] TOZAWA T., YAMANE Y., MUKAIYAMA T., "Diastereo- and Enantioselective Tandem Michael Addition and Lactonization between Silyl Enolates and α,β-Unsaturated Ketones Catalyzed by a Chiral Quaternary Ammonium Phenoxide", *Chemical Letters*, vol. 35, p. 360, 2006.

[TOZ 07] TOZAWA T., NAGAO H., YAMANE Y. *et al.*, "Enantioselective Synthesis of 3,4-Dihydropyran-2-ones by Domino Michael Addition and Lactonization with New Asymmetric Organocatalysts: Cinchona-Alkaloid-Derived Chiral Quaternary Ammonium Phenoxides", *Chemistry: An Asian Journal*, vol. 2, p. 123, 2007.

[URA 08] URAGUCHI D., KOSHIMOTO K., OOI T., "Chiral Ammonium Betaines: A Bifunctional Organic Base Catalyst for Asymmetric Mannich-Type Reaction of α-Nitrocarboxylates", *Journal of the American Chemical Society*, vol. 130, p. 10878, 2008.

[URA 10a] URAGUCHI D., KOSHIMOTO K., MIYAKE S. *et al.*, "Chiral Ammonium Betaines as Ionic Nucleophilic Catalysts", *Angewandte Chemie International Edition*, vol. 49, p. 5567, 2010.

[URA 10b] URAGUCHI D., KOSHIMOTO K., OOI T., "Flexible synthesis, structural determination, and synthetic application of a new C1-symmetric chiral ammonium betaine", *Chemical Communications*, vol. 46, p. 300, 2010.

[URA 10c] URAGUCHI D., KOSHIMOTO K., SANADA C. *et al.*, "Performance of C1-symmetric chiral ammonium betaines as catalysts for the enantioselective Mannich-type reaction of α-nitrocarboxylates", *Tetrahedron: Asymmetry*, vol. 21, p. 1189, 2010.

[URA 12a] URAGUCHI D., KOSHIMOTO K., OOI T., "Ionic Nucleophilic Catalysis of Chiral Ammonium Betaines for Highly Stereoselective Aldol Reaction from Oxindole-Derived Vinylic Carbonates", *Journal of the American Chemical Society*, vol. 134, p. 6972, 2012.

[URA 12b] URAGUCHI D., OYAIZU K., OOI T., "Nitroolefins as a Nucleophilic Component for Highly Stereoselective Aza Henry Reaction under the Catalysis of Chiral Ammonium Betaines", *Chemistry: A European Journal*, vol. 18, p. 8306, 2012.

[URA 15] URAGUCHI D., OYAIZU K., NOGUCHI H. *et al.*, "Chiral Ammonium Betaine-Catalyzed Highly Stereoselective Aza-Henry Reaction of α-Aryl Nitromethanes with Aromatic N-Boc Imines", *Chemistry: An Asian Journal*, vol. 10, p. 334, 2015.

[WAN 07a] WANG X., KITAMURA M., MARUOKA K., "New, Chiral Phase Transfer Catalysts for Effecting Asymmetric Conjugate Additions of α-Alkyl-α-cyanoacetates to Acetylenic Esters", *Journal of the American Chemical Society*, vol. 129, p. 1038, 2007.

[WAN 07b] WANG Y., YE J., LIANG X., "Convenient preparation of chiral α,β-epoxy ketones via Claisen-Schmidt condensation-epoxidation sequence", *Advanced Synthesis & Catalysis*, vol. 349, p. 1033, 2007.

[WAN 10] WANG X., LAN Q., SHIRAKAWA S. *et al.*, "Chiral bifunctional phase transfer catalysts for asymmetric fluorination of β-keto esters", *Chemical Communications*, vol. 46, p. 321, 2010.

[WAN 11] WANG L., SHIRAKAWA S., MARUOKA K., "Asymmetric Neutral Amination of Nitroolefins Catalyzed by Chiral Bifunctional Ammonium Salts in Water-Rich Biphasic Solvent", *Angewandte Chemie International Edition*, vol. 50, p. 5327, 2011.

[WAN 13a] WANG H.-Y., CHAI Z., ZHAO G., "Novel bifunctional thiourea–ammonium salt catalysts derived from amino acids: application to highly enantio- and diastereoselective aza-Henry reaction", *Tetrahedron*, vol. 69, p. 5104, 2013.

[WAN 13b] WANG H.-Y., ZHANG J.-X., CAO D.-D. *et al.*, "Enantioselective Addition of Thiols to Imines Catalyzed by Thiourea–Quaternary Ammonium Salts", *ACS Catalysis*, vol. 3, p. 2218, 2013.

[WAN 13c] WANG X., YANG T., CHENG X. *et al.*, "Enantioselective Electrophilic Trifluoromethylthiolation of β-Ketoesters: A Case of Reactivity and Selectivity Bias for Organocatalysis", *Angewandte Chemie International Edition*, vol. 52, p. 12860, 2013.

[WAN 14] WANG B., LIU Y., SUN C. *et al.*, "Asymmetric Phase-Transfer Catalysts Bearing Multiple Hydrogen-Bonding Donors: Highly Efficient Catalysts for Enantio- and Diastereoselective Nitro-Mannich Reaction of Amidosulfones", *Organic Letters*, vol. 16, p. 6432, 2014.

[WAN 15a] WANG B., HE Y., FU X. *et al.*, "A New Class of Squaramide-Containing Phase-Transfer Catalysts: Application to Asymmetric Fluorination of β-Keto Esters", *Synlett*, vol. 26, p. 2588, 2015.

[WAN 15b] WANG C., ZONG L., TAN C.-H., "Enantioselective Oxidation of Alkenes with Potassium Permanganate Catalyzed by Chiral Dicationic Bisguanidinium", *Journal of the American Chemical Society*, vol. 137, p. 10677, 2015.

[WAS 12] WASER M., GRATZER K., HERCHL R. *et al.*, "Design, synthesis, and application of tartaric acid derived N-spiro quaternary ammonium salts as chiral phase-transfer catalysts", *Organic & Biomolecular Chemistry*, vol. 10, p. 251, 2012.

[WEI 12a] WEI Y., HE W., LIU Y., *et al.*, "Highly Enantioselective Nitro-Mannich Reaction Catalyzed by Cinchona Alkaloids and N-Benzotriazole Derived Ammonium Salts", *Organic Letters*, vol. 14, p. 704, 2012.

[WEI 12b] WEIß M., BORCHERT S., RÉMOND E. *et al.*, "Asymmetric addition of a nitrogen nucleophile to an enoate in the presence of a chiral phase-transfer catalyst: A novel approach toward enantiomerically enriched protected β-amino acids", *Heteroatom Chemistry*, vol. 23, p. 202, 2012.

[WLA 99] WLADISLAW B., MARZORATI L., BIAGGIO F.C. *et al.*, "Sulfenylation of β-keto sulfoxides. III. Diastereoselectivity induced by a chiral phase transfer catalyst", *Tetrahedron*, vol. 55, pp. 12023–12030, 1999.

[WOZ 15] WOŹNIAK Ł., MURPHY J.J., MELCHIORRE P., "Photo-organocatalytic Enantioselective Perfluoroalkylation of β-Ketoesters", *Journal of the American Chemical Society*, vol. 137, p. 5678, 2015.

[WU 13] WU S., PAN D., CAO C. *et al.*, "Diastereoselective and Enantioselective Epoxidation of Acyclic β-Trifluoromethyl-β,β-Disubstituted Enones by Hydrogen Peroxide with a Pentafluorinated Quinidine-Derived Phase-Transfer Catalyst", *Advanced Synthesis & Catalysis*, vol. 355, p. 1917, 2013.

[XIA 15] XIAOYANG D., PEI X., ZHIYU W. *et al.*, "Asymmetric Sulfenylation of Glycine Derivative Catalyzed", *Austin Journal of Analytical and Pharmaceutical Chemistry*, vol. 2, p. 1032, 2015.

[YAN 03] YANG H.-M., WU H.-S., "Interfacial Mechanism and Kinetics of Phase-Transfer Catalysis", *Catalysis Reviews*, vol. 45, p. 463, 2003.

[YAN 12] YANG Y., MOINODEEN F., CHIN W. *et al.*, "Pentanidium–Catalyzed Enantioselective α-Hydroxylation of Oxindoles Using Molecular Oxygen", *Organic Letters*, vol. 14, p. 4762, 2012.

[YAO 12] YAO H., LIAN M., LI Z. *et al.*, "Asymmetric Direct α-Hydroxylation of β-Oxo Esters Catalyzed by Chiral Quaternary Ammonium Salts Derived from Cinchona Alkaloids", *Journal of Organic Chemistry*, vol. 77, p. 9601, 2012.

[YEB 11] YEBOAH E.M.O., YEBOAH S.O., SINGH G.S., "Recent applications of Cinchona alkaloids and their derivatives as catalysts in metal-free asymmetric synthesis", *Tetrahedron*, vol. 67, p. 1725, 2011.

[YOO 10] YOO M.-S., KIM D.-G., HA M.W. *et al.* "Synthesis of (αR,βS)-epoxyketones by asymmetric epoxidation of chalcones with cinchona phase-transfer catalysts", *Tetrahedron Letters*, vol. 51, p. 5601, 2010.

[ZHA 00] ZHANG F.-Y., COREY E.J., "Highly Enantioselective Michael Reactions Catalyzed by a Chiral Quaternary Ammonium Salt. Illustration by Asymmetric Syntheses of (S)-Ornithine and Chiral 2-Cyclohexenones", *Organic Letters*, vol. 2, p. 1097, 2000.

[ZHA 07] ZHAO H., QIN B., LIU X. *et al.*, "Enantioselective trifluoromethylation of aromatic aldehydes catalyzed by combinatorial catalysts", *Tetrahedron*, vol. 63, p. 6822, 2007.

[ZHA 12] ZHANG W.-Q., CHENG L.-F., YU J. *et al.*, "A Chiral Bis(betaine) Catalyst for the Mannich Reaction of Azlactones and Aliphatic Imines", *Angewandte Chemie International Edition*, vol. 51, p. 4085, 2012.

[ZON 14] ZONG L., BAN X., KEE C.W. *et al.*, "Catalytic Enantioselective Alkylation of Sulfenate Anions to Chiral Heterocyclic Sulfoxides Using Halogenated Pentanidium Salts", *Angewandte Chemie International Edition*, vol. 53, p. 11849, 2014.

Index

A, B, C

aldehyde, 1–22, 28, 30, 33–35, 38–55, 109, 110, 114, 139, 140, 142, 150, 152
amine, 2, 3, 6, 23, 31, 35, 36
betaine, 90, 111, 124, 133, 135, 136, 140, 141, 145, 147–150, 152
Brønsted
 acid, 1, 62, 65–68, 149
 base catalysis, 122
chiral
 molecules, 27
 quaternary ammonium salt, 87, 89, 100, 116–121, 142, 152
chirality, 90, 98, 102, 125, 132
cooperative catalysis, 27, 91

D, E

deprotonation, 54, 56, 65, 69, 91, 93, 94, 96, 103, 105, 111, 123, 124, 130, 135, 146, 147
electron transfer, 23, 28, 37, 45, 63, 67
enamine, 1–18, 28–39, 42–46, 48, 50, 52–58, 61, 72
enantioselective synthesis, 22, 122, 125

H, I, L

hydrogen bonding, 27, 62–65, 67, 92, 97, 110, 130
iminium, 1, 2, 6, 20, 31, 40, 45, 50, 53, 57–61
ion-pairing, 1, 69
Lewis base catalysis, 90, 135, 136, 137, 139, 140, 142, 143, 145

N, O, P

nucleophilic addition, 20
oxygenation, 116, 117
phase-transfer catalysis, 87
phenolate, 120, 135
photocatalysis, 24, 27, 28, 62, 65, 70

R, V

radical chain, 26, 31, 32, 34, 36, 40, 43, 46–49, 54, 69, 70, 132
reduction, 24, 30, 36, 40, 43, 45, 47, 48, 58, 67, 68, 71, 77, 155
visible light, 23–37, 34, 37, 38, 64, 65, 70, 72

Printed in the United States
By Bookmasters